REALITY IS BROKEN

Jane McGonigal, PhD. is a world-renowned creator of alternate reality games and the Director of Games Research and Development at the Institute of the Future, a non-profit research group based in Palo Alto, California. She has advised companies such as Microsoft, McDonald's, Intel and Disney, and created games for organisations such as the World Bank. *Business Week* has named her one of its "Top Ten Innovators to Watch" and in 2010 Oprah Winfrey listed her as one of the ten most inspiring women in the world.

'Jane McGonigal's groundbreaking research offers a surprising solution to how we can build stronger communities and collaborate at extreme scales: by playing bigger and better games. And no one knows more about how to design world-changing games than McGonigal. *Reality Is Broken* is essential reading for anyone who wants to play a hand in inventing a better future'
Jimmy Wales, founder of Wikipedia

'McGonigal is persuasive and precise in explaining how games can transform our approach to those things we know we should do . . . she is also adept at showing how good games expose the alarming insubstantiality of much everyday experience. McGonigal is a passionate advocate . . . Given the power and the darker potentials of the tools she describes, we must hope that the world is listening'
Observer

'An intensely optimistic book . . . intriguing and thought-provoking' *New Statesman*

'*Reality Is Broken* is a compelling exploration of why playing games makes us feel so good, and why, far from being a distraction from reality, technology-led games are increasingly providing solutions to our daily dissatisfactions . . . If the world of gaming seems alien to you, this book will crack it wide open. For experienced gamers, it will likely inspire you to play—or even invent—better, more meaningful games. Despite her expertise, McGonigal's book is never overly technical, and as with a good computer game, anyone, regardless of gaming experience, is likely to get sucked in'
New Scientist

'Forget everything you know, or think you know, about online gaming. Like a blast of fresh air, *Reality Is Broken* blows away the tired stereotypes and reminds us that the human instinct to play can be harnessed for the greater good. With a stirring blend of energy, wisdom, and idealism, Jane McGonigal shows us how to start saving the world one game at a time'

Carl Honoré, author of *In Praise of Slowness* and *Under Pressure*

JANE McGONIGAL

Reality is Broken

Why Games Make Us Better and
How They Can Change the World

VINTAGE BOOKS
London

Published by Vintage 2012

10 9

Copyright © Jane McGonigal 2011

Jane McGonigal has asserted her right under the Copyright, Designs
and Patents Act 1988 to be identified as the author of this work

Excerpt from *The Grasshopper: Games, Life and Utopia* by
Bernard Suits. Reprinted by permission of Broadview Press.

First published in Great Britain in 2011 by
Jonathan Cape

Vintage
Random House, 20 Vauxhall Bridge Road,
London SW1V 2SA

www.vintage-books.co.uk

Addresses for companies within The Random House Group Limited
can be found at: www.randomhouse.co.uk/offices.htm

The Random House Group Limited Reg. No. 954009

A CIP catalogue record for this book
is available from the British Library

ISBN 9780099540281

Penguin Random House is committed to a sustainable future for
our business, our readers and our planet. This book is made from
Forest Stewardship Council® certified paper.

Printed and bound in Great Britain by Clays Ltd, Elcograf S.p.A.

for my husband, Kiyash,

who is better at every game than I am,

except for Werewolf

It is games that give us something to do when there is nothing to do. We thus call games "pastimes" and regard them as trifling fillers of the interstices of our lives. But they are much more important than that. They are clues to the future. And their serious cultivation now is perhaps our only salvation.

—BERNARD SUITS, philosopher[1]

Contents

Reality Is Broken

Reality Is Broken

*Anyone who sees a hurricane coming should warn others.
I see a hurricane coming.*

*Over the next generation or two, ever larger numbers of
people, hundreds of millions, will become immersed in virtual worlds and online games. While we are playing, things
we used to do on the outside, in "reality," won't be happening anymore, or won't be happening in the same way. You
can't pull millions of person-hours out of a society without creating an atmospheric-level event.*

*If it happens in a generation, I think the twenty-first
century will see a social cataclysm larger than that caused
by cars, radios, and TV, combined. . . . The exodus of these
people from the real world, from our normal daily life, will
create a change in social climate that makes global warm-
ing look like a tempest in a teacup.*

— EDWARD CASTRONOVA,
Exodus to the Virtual World[1]

G amers have had enough of reality.

They are abandoning it in droves—a few hours here, an entire weekend there, sometimes every spare minute of every day for stretches at a time—in favor of simulated environments and online games. Maybe you are one of these gamers. If not, then you definitely know some of them.

Who are they? They are the nine-to-fivers who come home and apply all of the smarts and talents that are underutilized at work to plan and coordinate complex raids and quests in massively multiplayer online games like *Final Fantasy XI* and the *Lineage* worlds. They're the music lovers who have invested hundreds of dollars on plastic *Rock Band* and *Guitar Hero* instruments and spent night after night rehearsing, in order to become virtuosos of video game performance.

They're the *World of Warcraft* fans who are so intent on mastering the challenges of their favorite game that, collectively, they've written a quarter of a million wiki articles about the fictional universe—creating a wiki resource nearly one-tenth the size of the entire Wikipedia.[2] They're the *Brain Age* and *Mario Kart* players who take handheld game consoles everywhere they go, sneaking in short puzzles, races, and minigames as often as possible, and as a result nearly eliminating mental downtime from their lives.

They're the United States troops stationed overseas who dedicate so many hours a week to burnishing their *Halo 3* in-game service record that earning virtual combat medals is widely known as the most popular activity for off-duty soldiers. They're the young adults in China who have spent so much play money, or "QQ coins," on magical swords and other powerful game objects that the People's Bank of China intervened to prevent the devaluation of the yuan, China's real-world currency.[3]

Most of all, they're the kids and teenagers worldwide who would rather spend hours in front of just about any computer game or video game than do anything else.

These gamers aren't rejecting reality entirely. They have jobs, goals, schoolwork, families, commitments, and real lives that they care about. But as they devote more and more of their free time to game worlds, the *real* world increasingly feels like it's missing something.

Gamers want to know: Where, in the real world, is that gamer sense of being fully alive, focused, and engaged in every moment? Where is the gamer feeling of power, heroic purpose, and community? Where are the bursts of exhilarating and creative game accomplishment? Where is the heart-expanding thrill of success and team victory? While gamers may experience these pleasures occasionally in their real lives, they experience them almost constantly when they're playing their favorite games.

The real world just doesn't offer up as easily the carefully designed pleasures, the thrilling challenges, and the powerful social bonding afforded by virtual environments. Reality doesn't motivate us as effectively. Reality isn't engineered to maximize our potential. Reality wasn't designed from the bottom up to make us happy.

And so, there is a growing perception in the gaming community:

Reality, compared to games, is broken.

In fact, it is more than a perception. It's a phenomenon. Economist Edward Castronova calls it a "mass exodus" to game spaces, and you can see it already happening in the numbers. Hundreds of millions of people worldwide are opting out of reality for larger and larger chunks of time. In the United States alone, there are 183 million *active gamers* (individuals who, in surveys, report that they play computer or video games "regularly"—on average, thirteen hours a week).[4] Globally, the online gamer community—including console, PC, and mobile phone gaming—counts more than 4 million gamers in the Middle East, 10 million in Russia, 105 million in India, 10 million in Vietnam, 10 million in Mexico, 13 million in Central and South America, 15 million in Australia, 17 million in South Korea, 100 million in Europe, and 200 million in China.[5]

Although a typical gamer plays for just an hour or two a day, there are now more than 6 million people in China who spend at least twenty-two hours a week gaming, the equivalent of a part-time job.[6] More than 10 million "hardcore" gamers in the United Kingdom, France, and Germany spend at least twenty hours a week playing.[7] And at the leading edge of this growth curve, more than 5 million "extreme" gamers in the United States play on average forty-five hours a week.[8]

With all of this play, we have turned digital games—for our computers, for our mobile phones, and for our home entertainment systems—into what is expected to be a $68 billion industry annually by the year 2012.[9] And we are creating a massive virtual silo of cognitive effort, emotional energy, and collective attention lavished on game worlds instead of on the real world.

The ever-skyrocketing amounts of time and money spent on games are being observed with alarm by some—concerned parents, teachers, and politicians—and eagerness by others—the many technology industries that expect to profit greatly from the game boom. Meanwhile, they are met with bewilderment and disdain by more than a few nongamers, who still make up nearly half of the U.S. population, although their numbers are rapidly decreasing. Many of them deem gaming a clear waste of time.

As we make these value judgments, hold moral debates over the addictive quality of games, and simultaneously rush to achieve massive industry expansion, a vital point is being missed. The fact that so many people of all ages, all over the world, are choosing to spend so much time in game worlds is a sign of something important, a truth that we urgently need to recognize.

The truth is this: in today's society, computer and video games are fulfilling *genuine human needs* that the real world is currently unable to satisfy. Games are providing rewards that reality is not. They are teaching and inspiring and engaging us in ways that reality is not. They are bringing us together in ways that reality is not.

And unless something dramatic happens to reverse the resulting exodus, we're fast on our way to becoming a society in which a substantial portion of our population devotes its greatest efforts to playing games, creates its best memories in game environments, and experiences its biggest successes in game worlds.

Maybe this sounds hard to believe. To a nongamer, this forecast might seem surreal, or like science fiction. Are huge swaths of civilization really disappearing into game worlds? Are we really rushing headlong into a future where the majority of us use games to satisfy many of our most important needs?

If so, it will not be the first time that such a mass exodus from reality to games has occurred. Indeed, the very first written history of human gameplay, Herodotus' *Histories*, the ancient Greek account of the Persian Wars—dating back more than three thousand years—describes a nearly identical scenario. While the oldest known game is the ancient counting game Mancala— evidence shows it was played during Egypt's age of empires, or the fifteenth to the eleventh centuries BC—it was not until Herodotus that anyone thought to record the origins or cultural functions of these games. And from his ancient text, we can learn a great deal about what's happening today—and what's almost certainly coming next.

It's a bit counterintuitive to think about the future in terms of the past. But as a research director at the Institute for the Future—a nonprofit think tank in Palo Alto, California, and the world's oldest future-forecasting organization— I've learned an important trick: to develop foresight, you need to practice hindsight. Technologies, cultures, and climates may change, but our basic human needs and desires—to survive, to care for our families, and to lead happy, purposeful lives—remain the same. So at IFTF we like to say, "To understand the future, you have to look back at least twice as far as you're looking ahead." Fortunately, when it comes to games, we can look even farther back than that. Games have been a fundamental part of human civilization for thousands of years.

In the opening book of *The Histories*, Herodotus writes:

> When Atys was king of Lydia in Asia Minor some three thousand years ago, a great scarcity threatened his realm. For a while people accepted their lot without complaining, in the hope that times of plenty would return. But when things failed to get better, the Lydians devised a strange remedy for their problem. The plan adopted against the famine was to engage in games one day so entirely as

not to feel any craving for food . . . and the next day to eat and abstain from games. In this way they passed eighteen years, and along the way they invented the dice, knuckle-bones, the ball, and all the games which are common.[10]

What do ancient dice made from sheep's knuckles have to do with the future of computer and video games? More than you might expect.

Herodotus invented history as we know it, and he has described the goal of history as uncovering moral problems and moral truths in the concrete data of experience. Whether Herodotus' story of an eighteen-year famine survived through gameplay is true or, as some modern historians believe, apocryphal, its moral truths reveal something important about the essence of games.

We often think of immersive gameplay as "escapist," a kind of passive retreat from reality. But through the lens of Herodotus' history, we can see how games could be a *purposeful* escape, a thoughtful and active escape, and most importantly an extremely helpful escape. For the Lydians, playing together as a nearly full-time activity would have been a behavior highly adaptive to difficult conditions. Games made life bearable. Games gave a starving population a feeling of power in a powerless situation, a sense of structure in a chaotic environment. Games gave them a better way to live when their circumstances were otherwise completely unsupportive and uninhabitable.

Make no mistake: we are no different from the ancient Lydians. Today, many of us are suffering from a vast and primal hunger. But it is not a hunger for food—it is a hunger for more and better engagement from the world around us.

Like the ancient Lydians, many gamers have already figured out how to use the immersive power of play to distract themselves from their hunger: a hunger for more satisfying work, for a stronger sense of community, and for a more engaging and meaningful life.

Collectively, the planet is now spending more than 3 billion hours a week gaming.

We are starving, and our games are feeding us.

———

AND SO, in 2011, we find ourselves at a major tipping point.

We can stay on the same course. We can keep feeding our appetites with games. And we can watch the game industry continue to create bigger, better, and more immersive virtual worlds that provide increasingly compelling alternatives to reality.

If we stay this course, we will almost certainly see the exodus from reality continue. Indeed, we are already well on our way to a world in which many of us, like the ancient Lydians, spend half our time gaming. Given all the problems in the world, would it really be so bad to pass the coming decades as the Lydians did?

Or we could try to reverse course. We could try to block gamers' exit from reality—perhaps by culturally shaming them into spending more time in reality, or by trying to keep video games out of the hands of kids, or, as some U.S. politicians have already proposed, by heavily taxing them so that gaming becomes an unaffordable lifestyle.[11]

To be honest, none of those options sounds like a future I'd want to live in.

Why would we want to waste the power of games on escapist entertainment?

Why would we want to waste the power of games by trying to squelch the phenomenon altogether?

Perhaps we should consider a third idea. Instead of teetering on the tipping point between games and reality, what if we threw ourselves off the scale and tried something else entirely?

What if we decided to use everything we know about game design to fix what's wrong with reality? What if we started to live our real lives like gamers, lead our real businesses and communities like game designers, and think about solving real-world problems like computer and video game theorists?

Imagine a near future in which most of the real world works more like a game. But is it even possible to create this future? Would it be a reality we would be happier to live in? Would it make the world a better place?

When I consider this potential future, it's not just a hypothetical idea. I've already posed it as a very real challenge to the one community who can truly help launch this transformation: the people who make games for a living. I'm one of them—I've been designing games professionally for the past decade. And I've come to believe that people who know how to make games need to start focusing on the task of making real life better for as many people as possible.

I haven't always been so sure of this mission. It has taken a good ten years of research and a series of increasingly ambitious game projects to get to this point.

Back in 2001, I started my career by working on the fringes of the game-design industry, at tiny start-up companies and experimental design labs. More often than not, I was working for free, designing puzzles and missions for low-budget computer and mobile phone games. I was happy when they were played by a few hundred people, or—when I was really lucky—a few thousand. I studied those players as closely as possible. I watched them while they played, and I interviewed them afterward. I was just starting to learn what gives games their power.

During those early years, I was also a "starving" graduate student—earning a PhD in performance studies from the University of California at Berkeley. I was the first in my department to study computer and video games, and I had to make it up as I went along, bringing together different findings from psychology, cognitive science, sociology, economics, political science, and performance theory in order to try to figure out exactly what makes a good game work. I was particularly interested in how games could change the way we think and act in everyday life—a question that, back then, few, if any, researchers were looking at.

Eventually, as a result of my research, I published several academic papers (and eventually a five-hundred-page dissertation) proposing how we could leverage the power of games to reinvent everything from government, health care, and education to traditional media, marketing, and entrepreneurship—even world peace. And increasingly, I found myself called on to help large companies and organizations adopt game design as an innovation strategy—from the World Bank, the American Heart Association, the National Academy of Sciences, and the U.S. Department of Defense to McDonald's, Intel, the Corporation for Public Broadcasting, and the International Olympic Committee. You'll read about many of the games I created with these organizations in this book—and for the first time, I'll be sharing my design motivations and strategies.

The inspiration for this book came in the spring of 2008, when I was invited to deliver the annual "rant" at the Game Developers Conference, the most important industry gathering of the year. The rant is supposed to be a wake-up call, a demand to shake up the industry. It's always one of the most popular sessions at the conference. That year, the room was packed to standing-room capacity with more than a thousand of the world's leading game designers and developers. And in my rant, they heard the same argument you're reading here: that reality is broken, and we need to start making games to fix it.

When I finished, the applause and cheers took what seemed like forever to die down. I had been nervous that my rant would be rejected by my peers. Instead, it seemed to strike a chord with the industry. I started to get e-mails every single day from people who had heard about the rant or read the transcript online and wanted to help. Some were just starting out in the industry and had no idea how to go about doing it. Others were industry leaders who genuinely wanted to change the direction of games for good. Seemingly overnight, start-up companies were founded, capital was raised, and today there are hundreds of games in development that aspire to change reality for the better. I wouldn't dream of taking credit for this turn of events, of course. I was just lucky enough to be one of the first people to see it happening, and one of the strongest voices cheering it on.

In 2009, I was invited back to the Game Developers Conference to give a keynote address about what game developers needed to do over the next decade to reinvent reality as we know it. This time, I wasn't surprised to discover that some of the most popular sessions at the conference were about "games for personal and social change," "positive impact games," "social reality games," "serious games," and "leveraging the play of the planet." Everywhere I turned, I saw evidence that this movement to harness the power of games for good had already started to happen. Suddenly, my personal mission to see a game developer win a Nobel Peace Prize in the next twenty-five years didn't seem so far-fetched.

When I look at the remarkable world-changing work game developers are starting to do, I see an opportunity to reinvent the ancient history of games for the twenty-first century.

Some twenty-five hundred years ago, Herodotus looked back and saw the early games played by the Greeks as an explicit attempt to alleviate suffering. Today, I look forward and I see a future in which games once again are explicitly designed to improve quality of life, to prevent suffering, and to create real, wide-spread happiness.

When Herodotus looked back, he saw games that were large-scale systems, designed to organize masses of people and make an entire civilization more resilient. I look forward to a future in which massively multiplayer games are once again designed in order to reorganize society in better ways, and to get seemingly miraculous things done.

Herodotus saw games as a surprising, inventive, and effective way to intervene in a social crisis. I, too, see games as potential solutions to our most pressing shared problems. He saw that games could tap into our strongest survival instincts. I see games that once again will confer evolutionary advantage on those who play them.

Herodotus tells us that in the past games were created as a virtual solution to unbearable hunger. And, yes, I see a future in which games continue to satisfy our hunger to be challenged and rewarded, to be creative and successful, to be social and part of something larger than ourselves. But I also see a future in which the games we play *stoke* our appetite for engagement, pushing and enabling us to make stronger connections—and bigger contributions—to the world around us.

The modern history of computer and video games is the story of game designers ascending to very powerful positions in society, effectively enthralling the hearts and minds—and directing the energies and attention—of increasingly large masses of people. Game designers today are extremely adept wielders of that power, no doubt more adept than any game designers in all of human history. They have been honing their craft and refining their tactics for thirty years now. And so it is that more and more people are being drawn to the power of computer and video games—and finding themselves engaged by them for longer and longer periods of time, for greater and greater stretches of their lives.

Amazingly, some people have no interest in understanding why this is happening or figuring out what we could do with it. They will never pick up a book about games, because they're already certain they know exactly what games are good for—wasting time, tuning out, and losing out on real life.

The people who continue to write off games will be at a major disadvantage in the coming years. Those who deem them unworthy of their time and attention won't know how to leverage the power of games in their communities, in their businesses, in their own lives. They will be less prepared to shape the future. And therefore they will miss some of the most promising opportunities we have to solve problems, create new experiences, and fix what's wrong with reality.

Fortunately, the gap between gamers and nongamers is growing smaller all the time. In the United States, the biggest gaming market in the world, the majority of us are already gamers. Some recent relevant statistics from the Entertainment Software Association's annual study of game players—the largest and most widely respected market research report of its kind:

- 69 percent of all heads of household play computer and video games.
- 97 percent of youth play computer and video games.
- 40 percent of all gamers are women.
- One out of four gamers is over the age of fifty.
- The average game player is thirty-five years old and has been playing for twelve years.
- Most gamers expect to continue playing games for the rest of their lives.[12]

Meanwhile, the scientific journal *Cyberpsychology, Behavior, and Social Networking* reported in 2009 that 61 percent of surveyed CEOs, CFOs, and other senior executives say they take daily game breaks at work.[13]

These numbers demonstrate how quickly a gaming culture can take hold. And trends from every continent—from Austria, Brazil, and the United Arab Emirates to Malaysia, Mexico, New Zealand, and South Africa—show that gamer markets are emerging rapidly with similarly diverse demographics. Over the next decade, these new markets will increasingly resemble, if not completely catch up to, those in leading gamer countries like South Korea, the United States, Japan, and the United Kingdom today.

As games journalist Rob Fahey famously pronounced in 2008: "It's inevitable: soon we will all be gamers."[14]

We have to start taking this growing gamer majority seriously. We are living in a world full of games and gamers. And so we need to decide now what kinds of games we should make together and how we will play them together. We need a plan for determining how games will impact our real societies and our real lives. We need a framework for making these decisions and for shaping these plans. This book, I hope, could serve as that framework. It's written for gamers and for everyone who will one day become a gamer—in other words, for virtually every person on this planet. It's an opportunity to understand now how games work, why humans are so drawn to them, and what they can do for us in our real lives.

If you are a gamer, it's time to get over any regret you might feel about spending so much time playing games. You have *not* been wasting your time. You have been building up a wealth of virtual experience that, as the first half of this book will show you, can teach you about your true self: what your core strengths are, what really motivates you, and what make you happiest. As you'll see, you have also developed world-changing ways of thinking, organizing, and acting. And, as this book reveals, there are already plenty of opportunities for you to start using them for real-world good.

If you don't have a lot of personal experience with games yet, then this book will help you jump-start your engagement with the most important medium of the twenty-first century. By the time you're finished reading it, you'll be deeply familiar with the most important games you can play today—and be able to imagine the kinds of important games we will make and play in the years to come.

If you're not already a gamer, it's entirely possible that you still might not become the kind of person to spend hours in front of a video game. But by reading this book, you will better understand the people who do. And even if you would *never* play computer or video games, let alone make one, you can benefit enormously from learning exactly how good games work—and how they can be used to fix real-world problems.

Game developers know better than anyone else how to inspire extreme effort and reward hard work. They know how to facilitate cooperation and collaboration at previously unimaginable scales. And they are continuously innovating new ways to motivate players to stick with harder challenges, for longer, and in much bigger groups. These crucial twenty-first-century skills can help all of us find new ways to make a deep and lasting impact on the world around us.

Game design isn't just a technological craft. It's a twenty-first-century way of thinking and leading. And gameplay isn't just a pastime. It's a twenty-first-century way of working together to accomplish real change.

Antoine de Saint Exupéry once wrote:

As for the future, your task is not to see it, but to enable it.

Games, in the twenty-first century, will be a primary platform for enabling the future.

SO LET ME describe the particular future that I want to create.

Instead of providing gamers with better and more immersive alternatives to reality, I want all of us to be responsible for providing the world at large with a better and more immersive reality. I want gaming to be something that everybody does, because they understand that games can be a real solution to problems and a real source of happiness. I want games to be something everybody learns how to design and develop, because they understand that games are a real platform for change and getting things done. And I want families, schools, companies, industries, cities, countries, and the whole world to come together to play them, because we're finally making games that tackle real dilemmas and improve real lives.

If we take everything game developers have learned about optimizing human experience and organizing collaborative communities and apply it to real life, I foresee games that make us wake up in the morning and feel thrilled to start our day. I foresee games that reduce our stress at work and dramatically increase our career satisfaction. I foresee games that fix our educational systems. I foresee games that treat depression, obesity, anxiety, and attention deficit disorder. I foresee games that help the elderly feel engaged and socially connected. I foresee games that raise rates of democratic participation. I foresee games that tackle global-scale problems like climate change and poverty. In short, I foresee games that augment our most essential human capabilities—to be happy, resilient, creative—and empower us to change the world in meaningful ways. Indeed, as you'll see in the pages ahead, such games are already coming into existence.

The future I've described here seems both desirable and plausible to me. But in order to create this future, several things need to happen.

We will have to overcome the lingering cultural bias against games, so that nearly half the world is not cut off from the power of games.

We need to build hybrid industries and unconventional partnerships, so that game researchers and game designers and game developers can work with engineers and architects and policy makers and executives of all kinds to harness the power of games.

Finally, but perhaps most importantly, we all need to develop our core game competencies so we can take an active role in changing our lives and enabling the future.

This book is designed to do just that. It will build up your ability to enjoy life more, to solve tougher problems, and to lead others in world-changing efforts.

In **Part I: Why Games Make Us Happy**, you'll go inside the minds of top game designers and game researchers. You'll find out exactly which emotions the most successful games are carefully engineered to provoke—and how these feelings can spill over, in positive and surprising ways, into our real lives and relationships.

In **Part II: Reinventing Reality**, you'll discover the world of alternate reality games. It's the rapidly growing field of new software, services, and experiences meant to make us as happy and successful in our real lives as we are when we're playing our favorite video games. If you've never heard of ARGs before, you may be shocked to discover how many people are already making and playing them. Hundreds of start-up companies and independent designers have devoted themselves to applying leading-edge game design and technologies to improving our everyday lives. And millions of gamers have already discovered the benefits of ARGs firsthand. In this section, you'll find out how ARGs are already starting to raise our quality of life at home and at school, in our neighborhoods and our workplaces.

Finally, in **Part III: How Very Big Games Can Change the World**, you'll get a glimpse of the future. You'll discover ten games designed to help ordinary people achieve the world's most urgent goals: curing cancer, stopping climate change, spreading peace, ending poverty. You'll find out how new participation platforms and collaboration environments are making it possible for anyone to help invent a better future, just by playing a game.

Ultimately, the people who understand the power and potential of games to both make us happy and change reality will be the people who invent our future. By the time you finish reading this book, *you* will be an expert on how good games work. With that knowledge, you'll make better choices about which games to play and when. More importantly, you'll be ready to start inventing your own new games. You'll be prepared to create powerful, alternate realities for yourself and for your family; for your school, your business, your neighborhood, or any other community you care about; for your favorite cause, for an entire industry, or for an entirely new movement.

We can play any games we want. We can create any future we can imagine.

Let the games begin.

PART ONE

Why Games Make Us Happy

⤳ *One way or another, if human evolution is to go on, we shall have to learn to enjoy life more thoroughly.*

—MIHÁLY CSÍKSZENTMIHÁLYI[1]

CHAPTER ONE

What Exactly Is a Game?

Almost all of us are biased against games today—even gamers. We can't help it. This bias is part of our culture, part of our language, and it's even woven into the way we use the words "game" and "player" in everyday conversation.

Consider the popular expression "gaming the system." If I say that you're gaming the system, what I mean is that you're exploiting it for your own personal gain. Sure, you're technically following the rules, but you're playing in ways you're not meant to play. Generally speaking, we don't admire this kind of behavior. Yet paradoxically, we often give people this advice: "You'd better start playing the game." What we mean is, just do whatever it takes to get ahead. When we talk about "playing the game" in this way, we're really talking about potentially abandoning our own morals and ethics in favor of someone else's rules.

Meanwhile, we frequently use the term "player" to describe someone who manipulates others to get what they want. We don't really trust players. We have to be on our guard around people who play games—and that's why we might warn someone, "Don't play games with me." We don't like to feel that someone is using strategy against us, or manipulating us for their personal amusement. We don't like to be played with. And when we say, "This isn't a game!," what we mean is that someone is behaving recklessly or not taking a situation seriously. This admonishment implies that games encourage and train people to act in ways that aren't appropriate for real life.

When you start to pay attention, you realize how collectively suspicious we are of games. Just by looking at the language we use, you can see we're wary of how games encourage us to act and who we are liable to become if we play them.

But these metaphors don't accurately reflect what it really means to play a well-designed game. They're just a reflection of our worst fears about games. And it turns out that what we're really afraid of isn't games; we're afraid of losing track of where the game ends and where reality begins.

If we're going to fix reality with games, we have to overcome this fear. We need to focus on how real games actually work, and how we act and interact when we're playing the same game *together*.

Let's start with a really good definition of *game*.

The Four Defining Traits of a Game

Games today come in more forms, platforms, and genres than at any other time in human history.

We have single-player, multiplayer, and massively multiplayer games. We have games you can play on your personal computer, your console, your handheld device, and your mobile phone—not to mention the games we still play on fields or on courts, with cards or on boards.

We can choose from among five-second minigames, ten-minute casual games, eight-hour action games, and role-playing games that go on endlessly twenty-four hours a day, three hundred sixty-five days a year. We can play story-based games, and games with no story. We can play games with and without scores. We can play games that challenge mostly our brains or mostly our bodies—and infinitely various combinations of the two.

And yet somehow, even with all these varieties, when we're playing a game, we just know it. There's something essentially unique about the way games structure experience.

When you strip away the genre differences and the technological complexities, all games share four defining traits: a *goal*, *rules*, a *feedback system*, and *voluntary participation*.

The **goal** is the specific outcome that players will work to achieve. It focuses their attention and continually orients their participation throughout the game. The goal provides players with *a sense of purpose*.

The **rules** place limitations on how players can achieve the goal. By removing or limiting the obvious ways of getting to the goal, the rules push players to explore previously uncharted possibility spaces. They *unleash creativity* and *foster strategic thinking*.

The **feedback system** tells players how close they are to achieving the goal. It can take the form of points, levels, a score, or a progress bar. Or, in its most basic form, the feedback system can be as simple as the players' knowledge of an objective outcome: "The game is over when . . ." Real-time feedback serves as a *promise* to the players that the goal is definitely achievable, and it provides *motivation* to keep playing.

Finally, **voluntary participation** requires that everyone who is playing the game knowingly and willingly accepts the goal, the rules, and the feedback. Knowingness *establishes common ground* for multiple people to play together. And the freedom to enter or leave a game at will ensures that intentionally stressful and challenging work is experienced as *safe* and *pleasurable* activity.

This definition may surprise you for what it lacks: interactivity, graphics, narrative, rewards, competition, virtual environments, or the idea of "winning"— all traits we often think of when it comes to games today. True, these are common features of many games, but they are not *defining* features. What defines a game are the goal, the rules, the feedback system, and voluntary participation. Everything else is an effort to reinforce and enhance these four core elements. A compelling story makes the goal more enticing. Complex scoring metrics make the feedback systems more motivating. Achievements and levels multiply the opportunities for experiencing success. Multiplayer and massively multiplayer experiences can make the prolonged play more unpredictable or more pleasurable. Immersive graphics, sounds, and 3D environments increase our ability to pay sustained attention to the work we're doing in the game. And algorithms that increase the game's difficulty as you play are just ways of redefining the goal and introducing more challenging rules.

Bernard Suits, the late, great philosopher, sums it all up in what I consider the single most convincing and useful definition of a game ever devised:

> Playing a game is the voluntary attempt to overcome unnecessary obstacles.[1]

That definition, in a nutshell, explains everything that is motivating and rewarding and fun about playing games. And it brings us to our first fix for reality:

↻ FIX #1: UNNECESSARY OBSTACLES

Compared with games, reality is too easy. Games challenge us with voluntary obstacles and help us put our personal strengths to better use.

To see how these four traits are essential to every game, let's put them to a quick test. Can these four criteria effectively describe what's so compelling about games as diverse as, say, golf, Scrabble, and *Tetris*?

Let's take golf to start. As a golfer, you have a clear goal: to get a ball in a series of very small holes, with fewer tries than anyone else. If you weren't playing a game, you'd achieve this goal the most efficient way possible: you'd walk right up to each hole and drop the ball in with your hand. What makes golf a game is that you willingly agree to stand really far away from each hole and swing at the ball with a club. Golf is engaging exactly because you, along with all the other players, have agreed to make the work more challenging than it has any reasonable right to be.

Add to that challenge a reliable feedback system—you have both the objective measurement of whether or not the ball makes it into the hole, plus the tally of how many strokes you've made—and you have a system that not only allows you to know when and if you've achieved the goal, but also holds out the hope of potentially achieving the goal in increasingly satisfying ways: in fewer strokes, or against more players.

Golf is, in fact, Bernard Suits' favorite, quintessential example of a game—it really is an elegant explanation of exactly how and why we get so thoroughly engaged when we play. But what about a game where the unnecessary obstacles are more subtle?

In Scrabble, your goal is to spell out long and interesting words with lettered tiles. You have a lot of freedom: you can spell any word found in the dictionary. In normal life, we have a name for this kind of activity: it's called typing. Scrabble turns typing into a game by restricting your freedom in several important ways. To start, you have only seven letters to work with at a time. You don't get to choose which keys, or letters, you can use. You also have to base your words on the words that other players have already created. And there's a finite number of times each letter can be used. Without these arbitrary limitations, I think we can all agree that spelling words with lettered tiles wouldn't be much of a game. Freedom to work in the most logical and efficient way possible is the very *opposite* of gameplay. But add a set of obstacles and a feedback system—in this case, points—that shows you exactly how well you're spelling long and complicated words in the face of these obstacles? You get a system of completely unnecessary work that has enthralled more than 150 million people in 121 countries over the past seventy years.

Both golf and Scrabble have a clear win condition, but the ability to win is not a necessary defining trait of games. *Tetris*, often dubbed "the greatest computer game of all time," is a perfect example of a game you cannot win.[2]

When you play a traditional 2D game of *Tetris*, your goal is to stack falling puzzle pieces, leaving as few gaps as possible in between them. The pieces fall faster and faster, and the game simply gets harder and harder. It never ends. Instead, it simply waits for you to fail. If you play *Tetris*, you are *guaranteed* to lose.[3]

On the face of it, this doesn't sound very fun. What's so compelling about working harder and harder until you lose? But in fact, *Tetris* is one of the most beloved computer games ever created—and the term "addictive" has probably been applied to *Tetris* more than to any single-player game ever designed. What makes *Tetris* so addictive, despite the impossibility of winning, is the intensity of the feedback it provides.

As you successfully lock in *Tetris* puzzle pieces, you get three kinds of feedback: *visual*—you can see row after row of pieces disappearing with a satisfying poof; *quantitative*—a prominently displayed score constantly ticks upward; and *qualitative*—you experience a steady increase in how challenging the game feels.

This variety and intensity of feedback is the most important difference between digital and nondigital games. In computer and video games, the interactive loop is satisfyingly tight. There seems to be no gap between your actions and the game's responses. You can literally see in the animations and count on the scoreboard your impact on the game world. You can also feel how extraordinarily attentive the game system is to your performance. It only gets harder when you're playing well, creating a perfect balance between hard challenge and achievability.

In other words, in a good computer or video game you're always playing on the very edge of your skill level, always on the brink of falling off. When you do fall off, you feel the urge to climb back on. That's because there is virtually nothing as engaging as this state of working at the very limits of your ability—or what both game designers and psychologists call "flow."[4] When you are in a state of flow, you want to stay there: both quitting *and* winning are equally unsatisfying outcomes.

The popularity of an unwinnable game like *Tetris* completely upends the stereotype that gamers are highly competitive people who care more about winning than anything else. Competition and winning are *not* defining traits of games—nor are they defining interests of the people who love to play them.

Many gamers would rather keep playing than win—thereby ending the game. In high-feedback games, the state of being intensely engaged may ultimately be more pleasurable than even the satisfaction of winning.

The philosopher James P. Carse once wrote that there are two kinds of games: *finite games*, which we play to win, and *infinite games*, which we play in order to keep playing as long as possible.[5] In the world of computer and video games, *Tetris* is an excellent example of an infinite game. We play *Tetris* for the simple purpose of continuing to play a good game.

LET'S TEST OUR proposed definition for a game with one final example, a significantly more complex video game: the single-player action/puzzle game *Portal*.

When *Portal* begins, you find yourself in a small, clinical-looking room with no obvious way out. There is very little in this 3D environment to interact with: a radio, a desk, and what appears to be a sleeping pod. You can shuffle around the tiny room and peer out the glass windows, but that's about

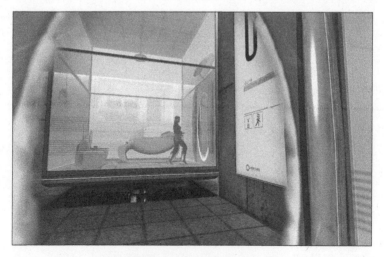

Screenshot from the first room of *Portal*.
(Valve Corporation, 2007)

it. There's nothing obvious to do: no enemies to fight, no treasure to pick up, no falling objects to avoid.

With so few clues for how to proceed, your goal at the start of the game is simply to figure out what your goals are. You might reasonably guess that your first goal is to get out of the sealed room, but you can't really be sure. It would seem that the main obstacle you face is that you have no idea what you're supposed to be doing. You're going to have to learn how to advance in this world on your own.

Well, not completely on your own. If you poke around the room enough, you might think to pick up a clipboard lying on the desk. This movement triggers an artificial intelligence system to wake up and start speaking to you. The AI informs you that you are about to undertake a series of laboratory tests. The AI does not tell you what you are being tested on. Again, it's up to you, the player, to figure it out.

What you eventually discover as you continue to play is that *Portal* is a game about escaping from rooms that operate according to rules you are un- aware of. You learn that each room is a puzzle, increasingly booby-trapped, and the game requires you to understand more and more complex physics in order to get out. If you don't teach yourself the physics of each new room— that is, if you don't learn the rules of the game—you'll be stuck there forever, listening to the AI system repeat herself.

Many, if not most, computer and video games today are structured this way. Players begin each game by tackling the obstacle of *not knowing what to do* and *not knowing how to play*. This kind of ambiguous play is markedly different from historical, predigital games. Traditionally, we have needed instructions in order to play a game. But now we're often invited to learn as we go. We explore the game space, and the computer code effectively constrains and guides us. We learn how to play by carefully observing what the game allows us to do and how it responds to our input. As a result, most gamers never read game manuals. In fact, it's a truism in the game industry that a well-designed game should be playable immediately, with no instruction whatsoever.

A game like *Portal* turns our definition of a game on its head, but doesn't destroy it. The four core elements of goals, rules, feedback, and voluntary participation remain the same—they just play out in a different order. It used to be that we were spoon-fed the goal and the rules, and we would then seek feedback on our progress. But increasingly, the feedback systems are what we learn first. They guide us toward the goal and help us decode the rules. And that's as powerful a motivation to play as any: discovering exactly what is possible in this brand-new virtual world.

I THINK it's fair to say that Suits' definition, and going forward *our* definition, holds up remarkably well against these diverse examples. Any well-designed game—digital or not—is an invitation to tackle an unnecessary obstacle.

When we understand games in this light, the dark metaphors we use for talking about games are revealed to be the irrational fears they really are. Gamers don't want to game the system. Gamers want to play the game. They want to explore and learn and improve. They're volunteering for unnecessary hard work—and they genuinely care about the outcome of their effort.

If the goal is truly compelling, and if the feedback is motivating enough, we will keep wrestling with the game's limitations—creatively, sincerely, and enthusiastically—for a very long time. We will play until we utterly exhaust our own abilities, or until we exhaust the challenge. And we will take the game seriously because there is nothing trivial about playing a good game. The game *matters*.

This is what it means to act like a gamer, or to be a truly *gameful* person. This is who we become when we play a good game.

But this definition leads us to a perplexing question. Why *on earth* are so many people volunteering to tackle such completely unnecessary obstacles? Why are we collectively spending 3 billion hours a week working at the very limits of our ability, for no obvious external reward? In other words: *Why do unnecessary obstacles make us happy?*

When it comes understanding how games really work, the answer to this question is as crucial as the four defining traits.

How Games Provoke Positive Emotion

Games make us happy because they are hard work that we choose for ourselves, and it turns out that almost nothing makes us happier than good, hard work.

We don't normally think of games as hard work. After all, we *play* games, and we've been taught to think of play as the very opposite of work. But nothing could be further from the truth. In fact, as Brian Sutton-Smith, a leading psychologist of play, once said, "The opposite of play isn't work. It's depression."[6]

When we're depressed, according to the clinical definition, we suffer from two things: a *pessimistic sense of inadequacy* and a *despondent lack of activity*. If we were to reverse these two traits, we'd get something like this: *an optimistic sense of our own capabilities* and *an invigorating rush of activity*. There's no clinical psychological term that describes this positive condition. But it's a perfect description of the emotional state of gameplay. A game is an opportunity to focus our energy, with relentless optimism, at something we're good at (or getting better at) and enjoy. In other words, *gameplay is the direct emotional opposite of depression*.

When we're playing a good game—when we're tackling unnecessary obstacles—we are actively moving ourselves toward the positive end of the emotional spectrum. We are intensely engaged, and this puts us in precisely the right frame of mind and physical condition to generate all kinds of positive emotions and experiences. All of the neurological and physiological systems that underlie happiness—our attention systems, our reward center, our motivation systems, our emotion and memory centers—are fully activated by gameplay.

This extreme emotional activation is the primary reason why today's most successful computer and video games are so addictive and mood-boosting. When we're in a concentrated state of optimistic engagement, it suddenly becomes biologically more possible for us to think positive thoughts, to make social connections, and to build personal strengths. We are actively conditioning our minds and bodies to be happier.

If only hard work in the real world had the same effect. In our real lives, hard work is too often something we do because we *have* to do it—to make a living, to get ahead, to meet someone else's expectations, or simply because someone else gave us a job to do. We resent that kind of work. It stresses us out. It takes time away from our friends and family. It comes with too much criticism. We're afraid of failing. We often don't get to see the direct impact of our efforts, so we rarely feel satisfied.

Or, worse, our real-world work isn't hard enough. We're bored out of our minds. We feel completely underutilized. We feel unappreciated. We are wasting our lives.

When we don't choose hard work for ourselves, it's usually not the right work, at the right time, for the right person. It's not perfectly customized for our strengths, we're not in control of the work flow, we don't have a clear picture of what we're contributing to, and we never see how it all pays off in the end. Hard work that someone else requires us to do just doesn't activate our happiness systems in the same way. It all too often doesn't absorb us, doesn't make us optimistic, and doesn't invigorate us.

What a boost to global net happiness it would be if we could positively activate the minds and bodies of hundreds of millions of people by offering them better hard work. We could offer them challenging, customizable missions and tasks, to do alone or with friends and family, whenever and wherever. We could provide them with vivid, real-time reports of the progress they're making and a clear view of the impact they're having on the world around them.

That's *exactly* what the game industry is doing today. It's fulfilling our need for better hard work—and helping us choose for ourselves the right work at the right time. So you can forget the old aphorism "All work and no play makes Jack a dull boy." All good gameplay *is* hard work. It's hard work that we enjoy and choose for ourselves. And when we do hard work that we care about, we are priming our minds for happiness.

The right hard work takes different forms at different times for different people. To meet these individual needs, games have been offering us increasingly diverse kinds of work for decades now.

There's **high-stakes work**, which is what many people think of first when it comes to video games. It's fast and action oriented, and it thrills us with the possibility not only of success but also of spectacular failure. Whether we're driving hairpin turns at top speeds in a racing video game like the *Gran Turismo* series or battling zombies in a first-person shooter game like *Left 4 Dead*, it's the risk of crashing, burning, or having our brains sucked out that makes us feel more alive.

But there's also **busywork,** which is completely predictable and monotonous. Busywork generally gets a bad rap in our real lives, but when we choose it for ourselves, it actually helps us feel quite contented and productive. When we're swapping multicolored jewels in a casual game like *Bejeweled* or harvesting virtual crops in a social game like FarmVille, we're happy just to keep our hands and mind occupied with focused activity that produces a clear result.

There's **mental work,** which revs up our cognitive faculties. It can be rapid-fire and condensed, like the thirty-second math problems in Nintendo's *Brain Age* games. Or it can be drawn-out and complex, like the simulated ten-thousand-year conquest campaigns in the real-time strategy game *Age of Empires*. Either way, we feel a rush of accomplishment when we put our brains to good use.

And then there's **physical work,** which makes our hearts beat faster, our lungs pump harder, our glands sweat like crazy. If the work is hard enough, we'll flood our brains with endorphins, the feel-good chemical. But more importantly, whether we're throwing punches in *Wii Boxing* or jumping around to *Dance Dance Revolution*, we just enjoy the process of getting ourselves completely worn out.

There's **discovery work,** which is all about the pleasure of actively investigating unfamiliar objects and spaces. Discovery work helps us feel confident, powerful, and motivated. When we're exploring mysterious 3D environments, like a vast city hidden in the sea in the role-playing shooter game *BioShock*, or when we're interacting with strange characters, like the fashionable undead teenagers who populate Tokyo in the handheld battle game *The World Ends with You*, we relish the chance to be curious about anything and everything.

Increasingly in computer and video games today there's **teamwork,** which emphasizes collaboration, cooperation, and contributions to a larger group. When we carve out special duties for ourselves in a complex mission like the twenty-five-player team raids in *World of Warcraft*, or when we're defending our friends' lives in a four-player cooperative game of the comic adventure *Castle Crashers*, we take great satisfaction in knowing we have a unique and important role to play in a much bigger effort.

Finally, there's **creative work.** When we do creative work, we get to make meaningful decisions and feel proud of something we've made. Creative work can take the form of designing our homes and families in the *Sims* games, or uploading video karaoke performances of ourselves to the *SingStar* network, or building and managing an online franchise in the *Madden NFL* games. For every creative effort we make, we feel more capable than when we started.

HIGH-STAKES WORK, busywork, mental work, physical work, discovery work, teamwork, and creative work—with all this hard work going on in our favorite games, I'm reminded of something the playwright Noël Coward once said: "Work is more fun than fun."

Sure, this sounds mildly absurd. Work more fun than fun? But when it comes to games, this is measurably and demonstrably true, thanks to a psychology research method known as "experience sampling."

Psychologists use the experience sampling method, or ESM, to find out how we really feel during different parts of our day. Subjects are interrupted at random intervals with a pager or by text message and asked to report two pieces of information: what they're doing and how they feel.[7] One of the most common findings of ESM research is that what we think is "fun" is actually mildly depressing.

Virtually every activity that we would describe as a "relaxing" kind of fun— watching television, eating chocolate, window-shopping, or just chilling out— doesn't make us feel better. In fact, we consistently report feeling worse afterward than when we started "having fun": less motivated, less confident, and less engaged overall.[8] But how can so many of us be so wrong about what's fun? Shouldn't we have a better intuitive sense of what actually makes us feel better?

We certainly have a strong intuitive sense of what makes us feel bad, and negative stress and anxiety are usually at the top of the list. ESM researchers believe that when we consciously seek out relaxing fun, we're usually trying to reverse these negative feelings. When we seek out passive entertainment and low-engagement activities, we're using them as a counterbalance to how stimulated and overwhelmed we feel.

But by trying to have easy fun, we actually often wind up moving ourselves too far in the opposite direction. We go from stress and anxiety straight to boredom and depression. We'd be much better off avoiding easy fun and seeking out *hard fun*, or hard work that we enjoy, instead.

Hard fun is what happens when we experience positive stress, or *eustress* (a combination of the Greek *eu*, for "well-being," and *stress*). From a physiological and a neurological standpoint, eustress is virtually identical to negative stress: we produce adrenaline, our reward circuitry is activated, and blood flow increases to the attention control centers of the brain. What's fundamentally different is our frame of mind.

When we're afraid of failure or danger, or when the pressure is coming from an external source, extreme neurochemical activation doesn't make us happy. It makes us angry and combative, or it makes us want to escape and shut down emotionally. It can also trigger avoidance behaviors, like eating, smoking, or taking drugs.[9]

But during *eustress*, we aren't experiencing fear or pessimism. We've generated the stressful situation on purpose, so we're confident and optimistic. When we choose our hard work, we enjoy the stimulation and activation. It makes us want to dive in, join together, and get things done. And this optimistic invigoration is way more mood-boosting than relaxing. As long as we feel capable of meeting the challenge, we report being highly motivated, extremely interested, and positively engaged by stressful situations. And these are the key emotional states that correspond with overall well-being and life satisfaction.

Hard fun leaves us feeling measurably better than when we started. So it's no surprise, then, that one of the activities for which ESM subjects report the highest levels of interest and positive moods both during *and* afterward is when they're playing games—including sports, card games, board games, and computer and video games.[10] The research proves what gamers already know:

within the limits of our own endurance, we would rather work hard than be entertained. Perhaps that's why gamers spend less time watching television than anyone else on the planet.[11]

As Harvard professor and happiness expert Tal Ben-Shahar puts it, "We're much happier *enlivening* time rather than killing time."[12]

THERE'S ONE MORE important emotional benefit to hard fun: it's called "fiero," and it's possibly the most primal emotional rush we can experience.

Fiero is the Italian word for "pride," and it's been adopted by game designers to describe an emotional high we don't have a good word for in English.[13] Fiero is what we feel after we triumph over adversity. You know it when you feel it—*and* when you see it. That's because we almost all express fiero in exactly the same way: we throw our arms over our head and yell.

The fact that virtually all humans physically express fiero in the same way is a sure sign that it's related to some of our most primal emotions. Our brains and bodies must have evolved to experience fiero early on the human timeline— and, in fact, neuroscientists consider it part of our "caveman wiring." Fiero, according to researchers at the Center for Interdisciplinary Brain Sciences Research at Stanford, is the emotion that first created a desire to leave the cave and conquer the world.[14] It's a craving for challenges that we can overcome, battles we can win, and dangers we can vanquish.

Scientists have recently documented that fiero is one of the most powerful neurochemical highs we can experience. It involves three different structures of the reward circuitry of the brain, including the mesocorticolimbic center, which is most typically associated with reward and addiction. Fiero is a rush unlike any other rush, and the more challenging the obstacle we overcome, the more intense the fiero.

A GOOD GAME is a unique way of structuring experience and provoking positive emotion. It is an extremely powerful tool for inspiring participation and motivating hard work. And when this tool is deployed on top of a network, it can inspire and motivate tens, hundreds, thousands, or millions of people at a time.

Anything else you think you know about games, forget it for now. All the good that comes out of games—every single way that games can make us happier in our everyday lives and help us change the world—stems from their ability to organize us around a voluntary obstacle.

Understanding that this is how games really work can help us stop worrying about how people might game our systems, and inspire us to start giving them real, well-designed games to play instead. If we actively surround ourselves with people playing the same game that we are, then we can stop being so wary of "players" playing their own game. When we know what it really means to play a good game, we can stop reminding each other: *This isn't a game*. We can start actively encouraging people instead: This *could* be a game.

CHAPTER TWO

The Rise of the Happiness Engineers

'm not the first person to notice that reality is broken compared with games, especially when it comes to giving us good, hard work. In fact, the science of happiness was first born thirty-five years ago, when an American psychologist by the name of Mihály Csíkszentmihályi observed the very same thing. In 1975, Csíkszentmihályi published a groundbreaking scientific study called *Beyond Boredom and Anxiety*. The focus of the study was a specific kind of happiness that Csíkszentmihályi named *flow*: "the satisfying, exhilarating feeling of creative accomplishment and heightened functioning."[1] He spent seven years researching this kind of intense, joyous engagement: when and where do we experience it most, and how can we create more of it?

Csíkszentmihályi (pronounced *cheek-SENT-me-high*) found a depressing lack of flow in everyday life, but an overwhelming abundance of it in games and gamelike activities. His favorite examples of flow-inducing activities were chess, basketball, rock climbing, and partner dancing: all challenging endeavors with a clear goal, well-established rules for action, and the potential for increased difficulty and improvement over time. Most importantly, flow activities were done for pure enjoyment rather than for status, money, or obligation.

During this kind of highly structured, self-motivated hard work, Csíkszentmihályi wrote, we regularly achieve the greatest form of happiness available to human beings: *intense, optimistic engagement with the world around us*. We feel fully alive, full of potential and purpose—in other words, we are completely activated as human beings.

Of course, it's possible to achieve this kind of extreme activation outside of games. But Csíkszentmihályi's research showed that flow was most *reliably* and most *efficiently* produced by the specific combination of self-chosen goals, personally optimized obstacles, and continuous feedback that make up the essential structure of gameplay. "Games are an obvious source of flow," he wrote, "and play is the flow experience *par excellence*."[2]

But if games are the most consistent and efficient source of joyous engagement in our lives, he wondered, then why did real life so infrequently resemble a game? Csíkszentmihályi argued that the failure of schools, offices, factories, and other everyday environments to provide flow was a serious moral issue, one of the most urgent problems facing humanity. Why should we needlessly spend the majority of our lives in boredom and anxiety, when games point to a clear and better alternative? "If we continue to ignore what makes us happy," he wrote, "we shall actively help perpetuate the dehumanizing forces which are gaining momentum day by day."

The solution seemed obvious to Csíkszentmihályi: create more happiness by structuring real work like game work. Games teach us how to create opportunities for freely chosen, challenging work that keeps us at the limits of our abilities, and those lessons can be transferred to real life. Our most pressing problems—depression, helplessness, social alienation, and the sense that nothing we do truly matters—could be effectively addressed by integrating more gameful work into our everyday lives.[3] It wouldn't be easy, he admitted. But if we failed to at least try to create more flow, we risked losing entire generations to depression and despair.

He ended his groundbreaking study by warning of two populations in greatest need of more gameful work: "Alienated children in the suburbs and bored housewives in the homes *need* to experience flow. If they cannot get it, they will find substitutes in the form of escape." This statement was eerily prophetic: today it is precisely these two demographic groups—suburban kids and women who are at home during the day—who spend the most time escaping into computer and video games.[4] Clearly, we haven't done enough to increase everyday flow.

Csíkszentmihályi was right about the need to reinvent reality to work more like a game. He was just too early. In 1975, the rest of the field of modern psychology was still largely focused on understanding mental illness and negative emotions, not optimal human experience. There wasn't enough critical momentum among his peers to pick up the problem of everyday happiness. Meanwhile, the tools we had in 1975 for inventing and sharing new games with mass audiences were still in their infancy. *Pong*, the first commercial video game, was just three years old. The Atari home console was still two years away from being released. And only one major research book had been published on the psychology of gameplay: a 1971 book titled, appropriately, *The Study of Games*.[5]

Today, however, we are in a very different position. Since Csíkszentmihályi's breakthrough study, two crucial things have happened, making it suddenly much more practical to improve quality of life with games: the rise of positive psychology and the explosion of the computer and video game industry.

Positive psychology is the relatively new field of science that studies "human flourishing," or how we achieve different kinds of happiness. For just over a decade now, positive-psychology researchers have been accumulating a formidable body of knowledge about how our brains and bodies work to help us achieve well-being and life satisfaction.

Meanwhile, the commercial game industry is putting all that knowledge to use. Game developers today understand that games become hits and make money in direct proportion to how much satisfaction they provide and how much positive emotion they provoke—in other words, how happy they make their players. As a result, game designers have been taught to relentlessly pursue happiness outcomes, including *flow*—and they've innovated a wide range of other happiness strategies along the way.

Happiness, of course, hasn't always been the explicit goal of the game industry, and not all game developers today share it. Plenty of game developers today still think more about fun and amusement than well-being and life satisfaction. But since the rise of positive psychology, the creative leaders of the industry have increasingly focused on the emotional and psychological impact of their games. More and more, the directors and designers of major game studios are drawing directly on research findings from positive psychology to make better games. The game industry has even produced a number of scientific research labs expressly devoted to investigating the neurobiology of gameplay emotions.

On the whole, a shift is clearly happening. As one journalist put it, the Microsoft game-testing lab "looks more like a psychological research institute than a game studio."[6] This is no accident. Game designers and developers are actively transforming what once was an intuitive art of optimizing human experience into an applied science. And as a result, they are becoming the most talented and powerful happiness engineers on the planet.

Today, these two historical trends—the science of happiness and the emotional evolution of the game industry—are intersecting. Thanks to positive psychologists, we know better than ever what kinds of experiences and activities really make us happy. And thanks to game developers, we have more and more powerful, and increasingly mobile, systems for providing intense, optimistic engagement and the emotional rewards we crave most.

That gives us our second fix for reality:

FIX #2: EMOTIONAL ACTIVATION

Compared with games, reality is depressing. Games focus our energy, with relentless optimism, on something we're good at and enjoy.

We are finally perfectly poised to harness the potential of games to make us happy and improve our everyday quality of life.

Let's take a look at how we got here.

IN 1983, a jazz pianist and sociologist named David Sudnow published a video game memoir, the first of its kind. It was a 161-page chronicle of his efforts to master one of the original home video games: the Atari ping-pong-style game *Breakout*.

Sudnow wasn't your stereotypical teen hanging out in an arcade. He was a forty-three-year-old professor with a successful side career in music and a full life by any objective measure. No one would have predicted that playing a video game would become more satisfying work for him than doing research or making music—least of all Sudnow himself. But to his great surprise, that's exactly what happened. For three months, Sudnow played *Breakout* as if it were his full-time job: "Fifty hours, a good five hours a day for ten days, in the afternoon, the evening, at three o'clock in the morning."

What was so captivating about *Breakout*? It was basically single-player *Pong*: you'd rotate a joystick knob to move a flat paddle along the bottom of the screen and wait for a falling ball to hit the paddle. Move the paddle, wait for the ball; move the paddle, wait for the ball. Your goal was to aim the ball using the paddle to knock bricks out of a wall at the top of the screen.

At first, this work was easy: your paddle was big, the ball fell slowly, and there were plenty of bricks to hit. But as you knocked out more and more bricks, the ball fell faster and bounced off the wall more erratically, and your paddle shrank to half its original size. It became increasingly difficult to keep the ball in play, and it took better and better aim to guide the ball toward the few remaining bricks. Once you missed five balls, the game was over.

As primitive a video game as it was, it nevertheless made for a perfect little voluntary workload. It had everything you'd want from an unnecessary ob-stacle: a clear goal (destroy a prison wall), arbitrary restrictions (use only a paddle and five balls), and instant feedback, both visual and audio (the bricks disappeared from the screen one at a time, always with a satisfying beep). The computer algorithms continuously adjusted the difficulty level to keep you playing at the very edge of your own abilities.

As Sudnow put it, "Here's all the motivation you'd ever want . . . and the prize seemed to be just holding on."[7] The game completely sustained his attention, even when he wasn't in front of the Atari console. "When I wasn't at the TV, I was practicing the sequence in my imagination, walking down the street, sitting in a café twirling a salt shaker, looking up during dinner in a Japanese restaurant at a bamboo and rice paper trellis with *Breakout*-like rectangles on the ceiling . . . just waiting to get back to the game."[8]

The better he got at the game, the more he wanted to play, and the more he played, the better he got. Sudnow was so taken aback by the intensity of this continuous feedback loop, he felt compelled to write an entire book to understand it. It's an extended, poetic meditation on the emotions of gameplay. He says almost everything we need to know about the emotional power of early video games in just these two famous sentences:

> This was a whole different business, nothing like I'd ever known, like night and day. . . . Thirty seconds of play, and I'm on a whole new plane of being, all my synapses wailing.[9]

What Sudnow describes is the extreme neurochemical activation that happens in our brains and bodies when we start to play a good computer or video game. He was intensely focused, highly motivated, creatively charged, and working at the very limits of his abilities. Immersion was almost instant. Flow was fast and virtually guaranteed.

From zero to peak experience in thirty seconds flat—no wonder video games caught on. Never before in human history could this kind of optimal, emotional activation be accessed so cheaply, so reliably, so quickly.

In the past, the deepest experiences of flow had required years of practice to achieve, or extraordinary settings. When Csíkszentmihályi first wrote about it, he was studying *expert* players of chess, or basketball, or rock climbing, or partner dancing. Flow was typically the result of years, if not decades, of learning the structure of an activity and strengthening the required skills and abilities. Otherwise, it required being immersed in a truly spectacular and unusual context, like dancing in the crowded streets during a carnival or skiing down an exceptionally challenging mountainside.

Flow wasn't supposed to come easy. But, as Sudnow and millions of other early gamers discovered, video games made it possible to experience flow almost immediately. Video games took the traditional properties of potentially flow-inducing activities—a goal, obstacles, increasing challenge, and voluntary participation—and then used a combination of direct physical input (the joystick), flexible difficulty adjustment (the computer algorithms), and instant visual feedback (the video graphics) to tighten the feedback loop of games dramatically. And this faster, tighter feedback loop allowed for more reliable hits of the emotional reward fiero: each microlevel of difficulty you survived prompted a split-second emotional high.

The result was a much faster cycle of learning and reward, and ultimately a sense of perfect and powerful control over a "microworld" on the screen. As Sudnow described, "The joystick-button box feels like a genuine implement of action. *Bam, bam, bam*, got you . . . Please don't miss, come on, do it, get that brick, easy does it, no surprises, now stay cool, don't panic, take it in stride, get it now. Get that closure. Video-game action. You know when you've got it like you know your first drunk."[10]

It was this fast, reliable fix of flow and fiero that kept Sudnow and all the earliest gamers coming back for more. It's no exaggeration to say that for many gamers, it probably felt like they had been waiting their whole lives for something like this: a seemingly free and endless supply of invigorating activity and every reason in the world to feel optimistic about their own abilities.

Then and now, faster, more reliable flow and fiero separates computer and video games from all the games that came before them. And that's what makes Sudnow's memoir of playing *Breakout* such an important historical artifact. He was the first to express what was so new and emotionally riveting about digital games, before the spectacular graphics, before the epic stories, and before the massively multiplayer worlds. Back then, *all* the emotional power of video games stemmed from the fact that they were interesting obstacles with better feedback and more adaptive challenges. As a result, they excelled at one thing and pretty much one thing only: provoking so much flow and fiero, they were nearly impossible to stop playing.

In fact, that was the whole point of these early video games: to keep playing as long as possible. One of the original and best-known Atari slogans was "Discover how far you can go." It was a constant battle just to stay in the game, but that was also the primary satisfaction of playing. Flow and fiero are the original rewards of video gameplay, and by playing against the tireless machine, we could endlessly produce them for ourselves.

Well, almost endlessly.

After three months, Sudnow finally exhausted his seemingly infinite source of flow. He played a perfect game of *Breakout*, destroying all the walls with just one ball. It was one of the biggest fiero moments of his life. But on that day, when he earned the highest possible score, his *Breakout* obsession ended. He had gone as far as flow and fiero could take him.

Fortunately for gamers, that's not the end of the story.

Flow, as positive psychologists have documented, is only one part of the overall happiness picture. It was the first kind of happiness these psychologists studied, but the science has progressed significantly since then. As Corey Lee M. Keyes, a psychology professor at Emory University, explains, "Flow is considered *part* of the science of happiness but not all. . . . It is more of a temporal state than a trait or condition of human functioning. While there are studies on how to prolong it, flow is not seen as something that people can live within all the time."[11]

Flow is exhilarating in the moment. It makes us feel energized. A major flow experience can improve our mood for hours, or even days, afterward. But because it's such a state of extreme engagement, it eventually uses up our physical and mental resources.

We can't sustain flow indefinitely—as much as we might want to. That's why, according to Keyes, human flourishing requires a more "continuous" approach to well-being. It can't just be all flow, all the time. We have to find ways to enjoy the world and relish life even when we're not operating at our peak human potential.

This is true in games as well as in life.

David Sudnow, it turns out, was so exhausted by the three months he spent suspended in nearly continuous *Breakout* flow that he subsequently stayed far away from video games for quite some time.

Too much flow can lead to happiness burnout. Meanwhile, too much fiero can lead to addiction—a word that Sudnow never once used in his memoir but which nevertheless inescapably leaps to mind. Fiero taps into some of our most primal hardwiring, and our emotional response can be extreme.

Recently, Allan Reiss, a professor of psychiatry and behavioral science at Stanford, led a team of researchers in a study of the neurochemistry of fiero in gamers. His laboratory captured MRIs of gamers' brains while they were playing particularly challenging video games. The researchers observed exceptionally intense activation of the addiction circuitry of the brain when gamers experienced moments of triumph. And as a result, the researchers identified fiero as the most likely underlying cause of why some gamers feel "addicted" to their favorite games.[12]

Gamer addiction is a subject the industry takes seriously—it's a frequent topic at industry conferences and on game developer forums: what causes gamer addiction, and how can you help your players avoid it? This might at first seem surprising: doesn't the industry want gamers to spend *more* time (and money) playing games, not less? And it's true: more gaming by more people *is* the primary goal of the industry. But the industry wants to create *lifelong gamers*: people who can balance their favorite games with full and active lives. And so we have what is perhaps the central dilemma of the game industry over the past thirty years: how to enable gamers to play more without diminishing their real lives. The industry knows that gamers crave flow and fiero—and the more game developers give it to them, the more time and money gamers will spend on their favorite games. But beyond a certain playing threshold—for most gamers, it seems to be somewhere around twenty hours a week—they start to wonder if they're perhaps missing out on real life.

Technology journalist Clive Thompson has a name for this phenomenon: *gamer regret*.[13] And he'll be the first to admit that he suffers from it as much as any other gamer. Thompson recalls checking his personal statistics one day—many games keep track of how many hours you've spent playing—and was shocked to see that he had clocked in thirty-six hours playing a single game in one week—as he described it, "a missing-time experience so vast one would normally require a UFO abduction to achieve it." He found himself vacillating between pride in what he'd accomplished in the virtual game environment and wondering if all that hard work had really been worth it.

As Thompson writes: "The dirty secret of gamers is that we wrestle with this dilemma all the time. We're often gripped by . . . a sudden, horrifying sense of emptiness when we muse on all the other things we could have done with our game time." He admits: "The elation I feel when I finish a game is always slightly tinged with a worrisome sense of hollowness. Wouldn't I have been better off doing something that was difficult and challenging *and* productive?"

This internal conflict plays out in discussion forums all over Web. The twin questions "How much time do you spend playing games?" and "How much time is too much?" are ubiquitous in the gaming community.

As a partial solution to gamer regret, many of the most addictive online games have implemented a "fatigue system." These systems are most commonly used in online games in South Korea and China, where the rates of online gaming for young men can average up to forty hours a week.[14] After three hours of consecutive online play, gamers receive 50 percent fewer rewards (and half the fiero) for accomplishing the same amount of work. After five hours, it becomes impossible to earn any rewards. In the United States, a softer touch is more commonly employed: *World of Warcraft* players, for example, accumulate "resting bonuses" for every hour they spend *not* playing the game. When they log back in, their avatar can earn up to double rewards until it's time to rest again.

But these measures are a stopgap at best. Trying to stop people from playing their favorite games will never work; motivated gamers hungry for flow and fiero will find a way around the restrictions and limitations. What's needed instead is for games to go beyond flow and fiero, which make us happy in the moment, to provide a more lasting kind of emotional reward. We need games that make us happier even when we're not playing. Only then will we find the right balance between playing our favorite games and making the most of our real lives.

Fortunately, that's exactly what's happening in the computer and video game market today. Games are increasingly teaching us the four secrets of how to make our own happiness—and they're giving us the power to make it anytime, anywhere.

The Four Secrets to Making Our Own Happiness

Many different competing theories of happiness have emerged from the field of positive psychology, but if there's one thing virtually all positive psychologists agree on, it's this: there are many ways to be happy, but we cannot *find* happiness. No object, no event, no outcome or life circumstance can deliver real happiness to us. We have to *make our own* happiness—by working hard at activities that provide their own reward.[15]

When we try to find happiness outside of ourselves, we're focused on what positive psychologists call "extrinsic" rewards—money, material goods, status, or praise. When we get what we want, we feel good. Unfortunately, the pleasures of found happiness don't last very long. We build up a tolerance for our favorite things and start to want more. It takes bigger and better rewards just to trigger the same level of satisfaction and pleasure. The more we try to "find" happiness, the harder it gets. Positive psychologists call this process "hedonic adaptation," and it's one of the biggest hindrances to long-term life satisfaction.[16] The more we consume, acquire, and elevate our status, the harder it is to stay happy. Whether it's money, grades, promotions, popularity, attention, or just plain material things we want, scientists agree: seeking out external rewards is a sure path to sabotaging our own happiness.

On the other hand, when we set out to make our own happiness, we're focused on activity that generates *intrinsic* rewards—the positive emotions, personal strengths, and social connections that we build by engaging intensely with the world around us. We're not looking for praise or payouts. The very act of what we're doing, the enjoyment of being fully engaged, is enough.

The scientific term for this kind of self-motivated, self-rewarding activity is *autotelic* (from the Greek words for "self," *auto*, and "goal," *telos*).[17] We do autotelic work because it engages us completely, and because intense engagement is the most pleasurable, satisfying, and meaningful emotional state we can experience.

As long as we are regularly immersed in self-rewarding hard work, we will be happy more often than not—no matter what else is going on in our lives. This is one of the earliest hypotheses of positive psychology, and a fairly radical idea. It contradicts what so many of us have been taught to believe—that we need life to be a certain way in order for us to be happy, and that the easier life is the happier we are. But the relationship between hard work, intrinsic reward, and lasting happiness has been verified and confirmed through hundreds of studies and experiments.

One well-known study conducted at the University of Rochester, published in 2009, neatly upturns one of the most common assumptions about how happiness works. Researchers tracked 150 recent college graduates for two years, monitoring their goals and reported happiness levels. They compared the rates at which the graduates achieved both extrinsic and intrinsic rewards, with self-reported levels of well-being and life satisfaction. The researchers' unequivocal conclusion: "The attainment of extrinsic, or 'American Dream,' goals—money, fame, and being considered physically attractive by others—*does not contribute to happiness at all.*" In fact, they reported, far from creating well-being, achieving extrinsic rewards "actually does contribute to some ill-being." If we let our desire for more and more extrinsic rewards monopolize our time and attention, it prevents us from engaging in autotelic activities that would actually increase our happiness.

On the other hand, in the same study the University of Rochester researchers found that individuals who focused on intrinsically rewarding activity, working hard to develop their personal strengths and build social relationships, for example, were measurably happier over the entire two-year period *completely regardless of external life circumstances* like salary or social status.

This research confirms what dozens of other major studies have found: happiness derived from intrinsic reward is incredibly resilient. Every time we engage in autotelic activities, the very opposite of hedonic adaptation occurs. We wean ourselves off consumption and acquisition as sources of pleasure and develop our *hedonic resilience*. As research psychologist Sonja Lyubomirsky, a leading expert on intrinsic reward, explains: "One of the chief reasons for the durability of happiness activities is that . . . they are hard won. You've devoted time and effort. . . . You have made these practices happen, and you have the ability to make them happen again. This sense of capability and responsibility is a powerful boost in and of itself." In other words, we become better able to protect and strengthen our quality of life, regardless of external circumstances. We rely less and less on unreliable and short-lived external rewards and take control of our own happiness. "When the source of positive emotion is yourself . . . , it can continue to yield pleasure and make you happy. When the source of positive emotion is yourself, it is *renewable*."[18]

The prevailing positive-psychology theory that we are the one and only source of our own happiness isn't just a metaphor. It's a biological fact. Our brains and bodies produce neurochemicals and physiological sensations that we experience, in different quantities and combinations, as pleasure, enjoyment, satisfaction, ecstasy, contentment, love, and every other kind of happiness. And positive psychologists have shown that we don't need to wait for life to trigger these chemicals and sensations for us. We can trigger them ourselves through scientifically measurable autotelic activities.

In fact, from a neurological and physiological point of view, "intrinsic reward" is really just another way of describing the emotional payoffs we get by stimulating our internal happiness systems.

By undertaking a difficult challenge, such as trying to finish a task in a shorter time than usual, we can produce in our own bodies a rush of adrenaline, the excitement hormone that makes us feel confident, energetic, and highly motivated.[19]

By accomplishing something that is very hard for us, like solving a puzzle or finishing a race, our brains release a potent cocktail of norepinephrine, epinephrine, and dopamine. These three neurochemicals in combination make us feel satisfied, proud, and highly aroused.[20]

When we make someone else laugh or smile, our brain is flooded with dopamine, the neurotransmitter associated with pleasure and reward. If we laugh or smile, too, the effect is even more pronounced.[21]

Every time we coordinate or synchronize our physical movements with others, such as in dance or sports, we release oxytocin into our bloodstream, a neurochemical that makes us feel blissed out and ecstatic.[22]

When we seek out what we might describe as "powerful" and "moving" stories, media, or live performances, we're actually triggering our vagus nerve, which makes us feel emotionally "choked up" in our chests and throats, or we're firing up our nervous system's pilomotor reflex, which gives us pleasurable chills and goose bumps.[23]

And if we provoke our curiosity by exposing ourselves to ambiguous visual stimulus, like a wrapped present or a door that is just barely ajar, we experience a rush of "interest" biochemicals also known as "internal opiates." These include endorphins, which make us feel powerful and in control, and beta-endorphin, a "well-being" neurotransmitter that is eighty times more powerful than morphine.

Few of us set out intentionally to trigger these systems. We don't think of happiness as a process of tapping strategically into our neurochemistry. We just know what feels good and meaningful and satisfying, and that's the kind of activity we'll do for its own sake.

Of course, we've also developed many external shortcuts to triggering our hardwired happiness systems: addictive drugs and alcohol, rich but unhealthy food, and chronic shopping, to name a few. But none of these methods are sustainable or effective in the long term. As scientists have shown, hedonic adaptation to extrinsic reward will cause our shortcut happiness behaviors to spiral out of control until they no longer work or we can no longer afford them, or even until they kill us.

Fortunately, we don't have to fight this losing battle. As long as we're focused on intrinsic and not extrinsic reward, we never run out of the raw materials for making our own happiness. We're hardwired with neurochemical systems to make all the happiness we need. We just have to work hard at things that activate us and immerse ourselves in challenging activities we enjoy for their own sake.

Writer and self-described happiness explorer Elizabeth Gilbert puts it best: "Happiness is the consequence of personal effort. . . . You have to participate relentlessly in the manifestations of your own blessings."[24] We have the biological capability to create our own happiness through hard work. And the harder we work to experience intrinsic rewards, the stronger our internal happiness-making capabilities become.

SO WHICH INTRINSIC rewards, exactly, are most essential to our happiness? There's no definitive list, but a few key ideas and examples appear over and over again in the scientific literature. My analysis of significant positive-psychology findings from the past decade suggests that intrinsic rewards fall into four major categories.[25]

First and foremost, we crave **satisfying work**, every single day. The exact nature of this "satisfying work" is different from person to person, but for everyone it means being immersed in clearly defined, demanding activities that allow us to see the direct impact of our efforts.

Second, we crave **the experience, or at least the hope, of being successful.** We want to feel powerful in our own lives and show off to others what we're good at. We want to be optimistic about our own chances for success, to aspire to something, and to feel like we're getting better over time.

Third, we crave **social connection**. Humans are extremely social creatures, and even the most introverted among us derive a large percentage of our happiness from spending time with the people we care about. We want to share experiences and build bonds, and we most often accomplish that by doing things that matter together.

Fourth, and finally, we crave **meaning**, or the chance to be a part of something larger than ourselves. We want to feel curiosity, awe, and wonder about things that unfold on epic scales. And most importantly, we want to belong to and contribute to something that has lasting significance beyond our own individual lives.

These four kinds of intrinsic rewards are the foundation for optimal human experience. They're the most powerful motivations we have other than our basic survival needs (food, safety, and sex). And what these rewards all have in common is that they're all ways of engaging deeply with the world around us—with our environment, with other people, and with causes and projects bigger than ourselves.

IF INTRINSIC REWARD is so much more satisfying and effective in boosting our happiness than extrinsic reward, then shouldn't we all naturally spend most of our time tackling unnecessary obstacles and engaging in autotelic activity?

Unfortunately, as Sonja Lyubomirsky eloquently explains: "We have been conditioned to believe that the wrong things will make us lastingly happy."[26] We've been sold the American dream. And increasingly, it's not just Americans who are giving up real happiness in favor of the pursuit of wealth, fame, and beauty. Thanks to the globalization of consumer and popular culture, everyone on the planet is being sold the same dream of extrinsic reward. This is especially true in emerging economies like China, India, and Brazil, where more and more people are being ushered onto the global hedonic treadmill, encouraged to consume more and to compete for limited natural resources as a way to increase their quality of life.

But there is cause for hope. One group is opting out of this soul-deadening, planet-exhausting hedonic grind, and in larger and larger numbers: hard-core gamers.

Games, after all, are the quintessential autotelic activity. We only ever play because we want to. Games don't fuel our appetite for extrinsic reward: they don't pay us, they don't advance our careers, and they don't help us accumulate luxury goods. Instead, games enrich us with intrinsic rewards. They actively engage us in satisfying work that we have the chance to be successful at. They give us a highly structured way to spend time and build bonds with people we like. And if we play a game long enough, with a big enough network of players, we feel a part of something bigger than ourselves—part of an epic story, an important project, or a global community.

Good games help us experience the four things we crave most—and they do it safely, cheaply, and reliably.

Good games *are* productive. They're producing a higher quality of life.

When we realize that this *reorientation toward intrinsic reward* is what's really behind the 3 billion hours a week we spend gaming globally, the mass exodus to game worlds is neither surprising nor particularly alarming. Instead, it's overwhelming confirmation of what positive psychologists have found in their scientific research: self-motivated, self-rewarding activity really does make us happier. More importantly, it's evidence that gamers aren't escaping their real lives by playing games.

They're actively making their real lives more rewarding.

CHAPTER THREE

More Satisfying Work

Playing *World of Warcraft* is such a satisfying job, gamers have collectively spent 5.93 million years doing it.

It sounds impossible, but it's true: if you add up all the hours that gamers across the globe have spent playing *World of Warcraft* since the massively multiplayer online (MMO) role-playing game (RPG) first launched in 2004, you get a grand total of just over 50 billion collective hours—or 5.93 million years.[1]

To put that number in perspective: 5.93 million years ago is almost exactly the moment in history that our earliest human ancestors first stood upright.[2] By that measure, we've spent as much time playing *World of Warcraft* as we've spent evolving as a species.

No other computer game has ever made so much money keeping so many players occupied for so long. Each *WoW* player spends on average between seventeen and twenty-two hours per week in the virtual world, more time than any other computer game attracts.[3] And the number of subscribers has steadily grown from 250,000 in January 2004 to more than 11.5 million in January 2010, making it the single largest paying game community in the world. (Like many MMORPGs, *WoW* requires players to pay a monthly fee—on average, fifteen dollars—in order to access the virtual world.) WoW developer Activision Blizzard currently reaps an estimated $5 million every single day in global subscription fees alone.[4]

What accounts for *World of Warcraft*'s unprecedented success? More than anything else, it's the feeling of "blissful productivity" that the game provokes.[5]

Blissful productivity is the sense of being deeply immersed in work that produces immediate and obvious results. The clearer the results, and the faster we achieve them, the more blissfully productive we feel. And no game gives us a better sense of getting work done than WoW.

Your primary job in *World of Warcraft* is self-improvement—a kind of work nearly all of us find naturally compelling. You have an avatar, and your job is to make that avatar better, stronger, and richer in as many different ways as possible: more experience, more abilities, stronger armor, more skills, more talent, and a bigger reputation.

Each of these improvable traits is displayed in your avatar profile, alongside a point value. You improve yourself by earning more points, which requires managing a constant work flow of quests, battles, and professional training. The more points you earn, the higher your level, and the higher your level, the more challenging work you unlock. This process is called "leveling up." The more challenging the work, the more motivated you are to do it, and the more points you earn . . . It's a *virtuous circle* of productivity. As Edward Castronova, who is a leading researcher of virtual worlds, puts it, "There is zero unemployment in *World of Warcraft*."[6] The WoW work flow is famously designed so that there is always something to do, always different ways to improve your avatar.

Some of the work is thrilling and high-stakes: it involves battling powerful opponents you're just barely strong enough to fight. Some of the work is exploratory: you're figuring out how to navigate around the many different regions of the kingdom, discovering new creatures and investigating strange environments. Some of the work is busywork: you study a virtual profession, like leatherworking or blacksmithing, and you collect and combine raw materials to help you ply your trade.

A lot of the work is teamwork: you can join forces with other players to take on quests that none of you is strong enough to survive alone, and you can go on raids that can only be completed by five, ten, or even twenty-five players working together. This kind of collaboration often involves strategic work before you take on the challenge. You have to figure out what role everyone will play in the raid, and you may have to rehearse and coordinate your actions many times to get it right.

Between the high-stakes work, the exploratory work, the busywork, the teamwork, and the strategic work, the hours of work definitely add up. There's so much to do, the typical WoW player puts in as many hours weekly as a part-time job. All in all, it takes the average WoW player a total of *five hundred hours* of gameplay to develop his or her avatar to the game's current maximum level, which is where many players say the fun *really* starts.[7] Now that's a labor of love.

So how exactly does a game convince a player to spend *five hundred hours* playing it just to get to the "fun" part?

For some players, it's the promise of ultimate challenge that makes the incredible workload worth it. At the highest levels of the game, you get to experience the extreme adrenaline rush of what players call the "endgame." Players who crave high-stakes work and extreme mental activation level up as fast as they can to reach the endgame, because that's where the most challenging opponents and the hardest work—in other words, the most invigorating, confidence-building gameplay—is available.

But there are plenty of online games that allow you to risk your virtual life and battle challenging opponents in adrenaline-producing environments—and you get to do it from the very start of the game. If that were the main reward for playing a game like *World of Warcraft*, the requirement of spending five hundred hours leveling up would be a bug, not a feature. The *process* of leveling up is easily as important, if not more important, than the endgame. As one player explains, "If all I wanted to do was run around and kill stuff, I could play *Counter-Strike* . . . and that game's free."[8] The players of WoW, and the many other subscription-based massively multiplayer online games like it, are paying for a particular privilege. They're paying for the privilege of higher in-game productivity.

Consider many fans' reactions when it was revealed that a highly anticipated new MMO, *Age of Conan*, would take just two hundred fifty hours of gameplay on average to reach the highest level. Bloggers described this as a "paltry" and "positively anemic" amount of work, and professional game critics worried that fans would reject an MMO that required "so little effort" to achieve the highest level.[9]

In real life, if someone gave you a task that normally took five hundred hours of work to finish, and then gave you a way to complete it in half that, you would probably be pretty pleased. But in game life, where the whole point for so many players is to get their hands on as much satisfying work as possible, two hundred fifty hours of work is a disappointment. For these dedicated MMO players, the possibility of reaching the highest level is simply justification for what they really love most: getting better.

No wonder Nick Yee, a leading researcher of MMOs and the first person to receive a PhD for studying WoW, has argued that the MMOs are really massively multiplayer work environments disguised as games. As Yee observes, "Computers were made to work for us, but video games have come to demand that we work for them." This is true—but, of course, *we* are really the ones who are asking to have more work. We want to be given more work—or rather, we want to be given more *satisfying* work.

This brings us to our next fix for reality:

○ FIX #3: MORE SATISFYING WORK

Compared with games, reality is unproductive. Games give us clearer missions and more satisfying, hands-on work.

Satisfying work always *starts* with two things: **a clear goal** and **actionable next steps** toward achieving that goal. Having a clear goal motivates us to act: we know what we're supposed to do. And actionable next steps ensure that we can make progress toward the goal immediately.

What if we have a clear goal, but we aren't sure how to go about achieving it? Then it's not work—it's a *problem*. Now, there's nothing wrong with having interesting problems to solve; it can be quite engaging. But it doesn't necessarily lead to satisfaction. In the absence of actionable steps, our motivation to solve a problem might not be enough to make real progress. Well-designed work, on the other hand, leaves no doubt that progress will be made. There is a *guarantee of productivity* built in, and that's what makes it so appealing.

WoW offers a guarantee of productivity with every quest you undertake. The world is populated by thousands of characters who are willing to give you special assignments—each one presented on an individual scroll that lists a clear goal, and why it matters, followed by actionable steps: where to go, step-by-step instructions for what to do when you get there, and a concrete measure of proof you're expected to gather to demonstrate your success. For example, here is an annotated version of a typical WoW quest:

> **QUEST: A Worthy Weapon**
> Bring the Blade of Drak'Mar to Jaelyne Evensong at the Argent Tournament Grounds. (*This is your goal.*)
>
> Of all the times to have such rotten luck! My tournament blade has gone missing and I need it for a match this afternoon. (*And this is why your goal matters.*)
>
> One of the bards tells me that travelers used to present winter hyacinths to a lonely maiden in return for gifts. Those hyacinths grow only on the ice flowing from the Ironwall Dam, on Crystalsong Forest's northwestern border with Icecrown. (*This is where to go.*)
>
> Gather the flowers and take them to Drak'Mar Lake in northeastern Dragonblight, near its border with Zul'Drak and Grizzly Hills. (*These are your step-by-step instructions.*)
>
> Return to Jaelyne Evensong in Icecrown and deliver the Blade. (*This is your proof of completion.*)

When you're on a WoW quest, there's never any doubt about what you're supposed to do, or where or how. It's not a game that emphasizes puzzle solving or trial-and-error investigation. You simply have to get the job done, and then you will collect your rewards.

Why do we crave this kind of guaranteed productivity? In *The How of Happiness*, Sonja Lyubomirsky writes that the fastest way to improve someone's everyday quality of life is to "bestow on a person a specific *goal*, something to do and to look forward."[10] When a clear goal is attached to a specific task, she explains, it gives us an energizing push, a sense of purpose.

That's why receiving more quests every time we complete one in *World of Warcraft* is more of a reward than the experience points and the gold we've earned. Each quest is another clear goal with actionable steps.

The real payoff for our work in WoW is to be rewarded with more opportunities for work. The design of the work flow is key here: the game constantly challenging you to try something *just a little bit* more difficult than what you've just accomplished. These microincreases in challenge are just big enough to keep sparking your interest and motivation — but never big enough to create anxiety or the sense of an ability gap. As one longtime *World of Warcraft* player explains, "When accepting a quest, you rarely have to question *if* you can complete it; you just need to figure out *when* you can fit it into your jam-packed hero schedule."[11] This endless series of goals and actionable steps is exactly what makes *World of Warcraft* so invigorating.

MOTIVATION AND REASONABLY assured progress: this is the start of satisfying work. But to be truly satisfied, we have to be able to *finish* our work as clearly as we started it. To finish work in a satisfying way, we must be able to see the results of our efforts as directly, immediately, and vividly as possible.

Visible results are satisfying because they mirror back to us a positive sense of our own capabilities. When we can see what we've accomplished, we build our sense of self-worth. As Martin Seligman, one of the founders of positive psychology, argues, "The most important resource-building human trait is productivity at work."[12] The key here is *resource building*: we like productive work because it makes us feel that we are developing our personal resources.

The famous heads-up display of *World of Warcraft*, which shows us our improvement in real time, is all about making our own resource building more visible. It constantly flashes positive feedback at players: +1 stamina, +1 intellect, +1 strength. We can count our own internal resources by these points, watching as we become more and more resourceful with every effort we make: able to inflict or sustain more damage, or able to cast more powerful spells.

We can also see the self-improving results of our game work just by looking at our avatar, which visibly bears more impressive armors, weapons, and jewels over time. And many players install a game modification that can show them a complete history of every quest they've completed—the ultimate, tangible record of work well done.

And it's not just self-improvement. At the highest levels of the game, during the most collaborative game missions, the raids, *collective* improvement is the focus. Players may join what are called "guilds," or long-term alliances with other players, to complete the most difficult raids. One popular WoW guide explains, "Raiding is about building and maintaining a team, a close-knit group of players who progress together."[13] As the guild's raid statistics and achievement statistics measurably improve, the satisfaction of resource building is amplified by celebrating it with so many others.

But perhaps the most compelling form of feedback we get from working in the *World of Warcraft* isn't strictly about us. It's a visual effect called *phasing*, which is designed to vividly show us our impact on the world around us.

This is how phasing works: When I play an MMO on my computer, most of the game content isn't on my hard drive. It's on a remote server that's processing the game experiences for me and thousands of other players at the same time. For the most part, if I'm in one part of the game world on my computer, and you're in the same part of the game world on your computer, and we're both playing on the same server, we see exactly the same world. The game server sends us exactly the same visual data about who's there, what they're doing, and what the environment looks like.

But in phasing, the server compares the game histories of different players in the environment and shows each player a different version of the world depending on what they've accomplished. When you complete a heroic quest or a high-level raid, your virtual world literally changes—you see different things from someone who hasn't finished the quest or raid. As one WoW FAQ explains: "Did you help a faction conquer an area? When you next return they'll have a camp set up with vendors and other services, and all the bad guys are gone! The same area now serves a different purpose reflecting your earlier work."[14]

It's a very powerful special effect. We're not only improving our characters; we're improving the whole world. As one player writes in an enthusiastic review of the phasing content, "Whether this is achieved though technical wizardry or just straight-up magic is unclear. Its integration is seamless, and it's incredibly satisfying. You feel like your actions are having a significant impact on the world around you."[15]

That is, after all, one of things we crave most in life. In his study *The Pleasures of Sorrow and Work*, Alain de Botton argues that work is "meaningful only when it proceeds briskly in the hands of a restricted number of actors and therefore where particular workers can make an imaginative connection between what they have done with their working days and their impact upon others."[16] In other words, we have to both be close enough to the action and see the results directly and quickly enough for work to satisfy our craving to make an impact. When we don't have visible results that we can clearly link to our own efforts, it is impossible to take real satisfaction in our work. Unfortunately, for many of us this is true of our everyday work lives.

In *Shop Class as Soul Craft*, author and motorcycle mechanic Matthew Crawford reflects on the psychological differences between manual labor and everyday office work. As he observes:

> Many of us do work that feels more surreal than real. Working in
> an office, you often find it difficult to see any tangible result from
> your efforts. What exactly have you accomplished at the end of
> any given day? Where the chain of cause and effect is opaque and

responsibility diffuse, the experience of individual agency can be elusive. . . . Is there a more "real" alternative?[17]

While it may not be the solution Crawford is referring to, games like World of Warcraft are just that: a more "real" alternative to the insubstantiality of so much everyday work. Although we think of computer games as virtual experiences, they do give us *real agency*: the opportunity to do something that feels concrete because it produces measurable results, and the power to act directly on the virtual world. And, of course, gamers are working with their hands, even if what they're manipulating is digital data and virtual objects. Until and unless the real work world changes for the better, games like WoW will fulfill a fundamental human need: the need to feel productive.

That's what it takes for work to satisfy us: it must present us with clear, immediately actionable goals as well as direct, vivid feedback. World of Warcraft does all of this brilliantly, and it does so *continuously*. As a result, every single day, gamers worldwide spend a collective 30 million hours working in World of Warcraft. With its thousands of potential quests, its ever-elusive endgame, and a server that generates more obstacles and opponents for you every time you log on, it is without a doubt one of the most satisfying work systems ever engineered. Even people who love their real jobs can be seduced by the blissful productivity it provokes in us—myself included.

The first time I sat down to play the game, my friend Brian cheerfully warned me that "World of Warcraft is the single most powerful IV drip of productivity ever created."

He wasn't kidding. That weekend, I spent twenty-four hours playing WoW—which was about twenty-three more hours than I'd intended.

What can I say? There was a *lot* of world-saving work to do.

Every time I completed a quest, I racked up experience points and gold. But more important than the points or treasure, from the moment I entered the online Kingdom of Azeroth, I was rich with *goals*. Every quest came with clear, urgent instructions—where to go, what to do, and why the fate of the kingdom hung in the balance of my getting it done as soon as possible.

When Monday morning came around, I resisted the idea of going back to "real" work. I knew this wasn't rational. But some part of me wanted to keep earning experience points, stacking up treasure, collecting my plus-ones, and checking off world-saving quests from my to-do list.

"Playing WoW just feels way more productive," I remember telling my husband.

I did go back to real work, of course. But it took me a while to shake the feeling that I'd rather be leveling up. Part of me felt like I was accomplishing more in the Kingdom of Azeroth than I was in my real life. And that's exactly the IV drip of productivity that *World of Warcraft* is so good at providing. It delivers a stream of work and reward as reliably as a morphine drip line.

When we play WoW, we get blissed out by our own productivity—and it doesn't matter that the work isn't real. The emotional rewards are real—and for gamers, that's what matters.

WORLD OF WARCRAFT is an example of *extreme-scale* satisfying work. Players commit to this work environment for extraordinary periods of time.

But there are also *microexamples* of games that generate the rewarding sense of capability and productivity. They're called "casual games," and they provide satisfying work in very quick bursts of productive play: as short as a few minutes to an hour. When interspersed with everyday work, they can be surprisingly boosting to everyday life satisfaction.

"Casual games" is an industry term for games that tend to be easy to learn, quick to play, and require far less computer memory and processing power than other computer and video games. (They're often played online in Web browsers, or on mobile phones.)

These games require less of a commitment than most video games: a casual game player might play his or her favorite game for just fifteen minutes a day, a few times a week.

Even if you don't consider yourself a gamer, you've probably played some casual games—including the versions of solitaire and *Minesweeper* that come preinstalled on so many computers. Other iconic casual games include *Bejeweled*, in which the goal is to rearrange brightly colored gems into sets of three;

Diner Dash, a simulation of being a waitress; and one of my own personal favorites, *Peggle,* which requires you to aim and shoot balls to knock out pegs from a kind of psychedelic pachinko board.

Most casual games are single-player, allowing gamers to sneak in a few minutes of play for themselves whenever they need it most. And one of the places we seem to need the boost of gaming most is, perhaps not surprisingly, at the office.

A recent major survey of high-level executives, including chief executive officers, chief financial officers, and presidents, revealed that 70 percent of them regularly play casual computer games while working. That's right: the vast majority of senior executives report taking daily computer game breaks that last on average between fifteen minutes and one hour.

How do these executives explain their tendency to play while working? Most of them say they turn to games to feel "less stressed out." This makes perfect sense—casual games are undoubtedly more effective than more passive ways of decreasing stress at work, like browsing the Web. By tackling an unnecessary obstacle in the middle of the workday, these executives are triggering a sense of self-motivation. They're shifting their mental awareness from the externally applied pressures of real work, or negative stress, to the internally generated pressure of game work, or positive stress. The executives reported feeling "more confident, more energetic, and more mentally focused" after playing a quick computer game—all hallmarks of eustress.

But even more interestingly, more than half of these gameful executives say they play during work in order "to feel more productive."[18] Now this is a statement that sounds crazy on the face of it—playing games to feel more productive at work? But this speaks to how much we all crave simple, hands-on work that feels genuinely productive. We turn to games to help us alleviate the frustrating sense that, in our real work, we're often not making any progress or impact.

As de Botton writes: "Long before we ever earned any money, we were aware of the necessity of keeping busy: we knew the satisfaction of stacking bricks, pouring water into and out of containers and moving sand from one pit to another, untroubled by the greater purpose of our actions."[19] In casual games, there is no greater purpose to our actions—we are simply enjoying our ability to make something happen.

WHETHER IT'S A SHORT, simple burst of video game productivity or entering into sprawling worlds designed to engage us in endless campaigns of satisfying activity, playing games can give us a taste of that elusive sense of individual agency and impact in a world where the work we do may be challenging, but our efforts often seem fruitless.

The best-designed game work feels more productive because it feels more *real*: the feedback comes stronger and faster, and the impact is more visible and vivid. And for many of us who aren't gratified enough by our day-to-day jobs or don't feel like our work is having a direct impact, gameful work is a real source of reward and satisfaction.

On the other hand, as gratifying as it is to rack up achievements and get the job done, it can be equally energizing—but in a very different way—to fail, fail, fail. This brings us to our next intrinsic reward, which is a kind of counterbalance to the experience of satisfying work. It's the *hope*—but not necessarily the achievement—of success.

CHAPTER FOUR

Fun Failure and
Better Odds of Success

No one likes to fail. So how is it that gamers can spend 80 percent of the time failing, and still love what they're doing?"

Games researcher Nicole Lazzaro likes to stump audiences with tough questions, and this is one of her favorites. Lazzaro, an expert on game-play emotions, has been working in the game industry for twenty years as a design consultant. She reports her research findings and design recommendations to the industry annually at the Game Developers Conference. And perhaps her most significant finding yet is this: gamers spend nearly all of their time failing. Roughly four times out of five, gamers don't complete the mission, run out of time, don't solve the puzzle, lose the fight, fail to improve their score, crash and burn, or die.[1]

Which makes you wonder: do gamers actually *enjoy* failing?

As it turns out, yes.

Lazzaro has long suspected that gamers love to fail, and a team of psychologists at the M.I.N.D. Lab in Helsinki, Finland, recently confirmed it with scientific evidence. When we're playing a well-designed game, failure doesn't disappoint us. It makes us happy in a very particular way: excited, interested, and most of all *optimistic*.[2]

If that finding surprises you, then you're not alone—the Finnish research-ers weren't expecting that result, either. But today, the "fun failure" study is considered one of the most important findings in the history of video game research.[3] It helped pinpoint for the first time exactly how a well-designed game helps players develop exceptional mental toughness.

Why Failure Makes Us Happy

The M.I.N.D. Lab is a state-of-the-art psychophysiology research center, packed with biometric systems designed to measure emotional response: heart rate monitors, brain wave monitors, electrical sensors, and more.

In 2005, to kick off a new research effort focused on emotional response to video games, the lab invited thirty-two gamers to play the highly popular *Super Monkey Ball 2* while hooked up to the biometric monitors. In the bowling-style game, players roll "monkey balls," or transparent bowling balls with mon-keys inside them, down crooked bowling lanes that happen to be floating in outer space. Throw a gutter ball at any point, and the monkey rolls right off the edge of the lane, whirling off into the atmosphere.

While the gamers played, the researchers measured three indicators of emo-tional engagement: heart rate, because we pump blood faster when we're emotion-ally aroused; skin conductivity, because we sweat more when we're under stress; and electrical activation of the facial muscles, because we move certain muscles like the zygomaticus major muscle, which pulls the corners of the mouth back and up into a smile, when we're happy.

After collecting all of this physiological data, the researchers compared it against a log of key gameplay events—just before rolling the monkey ball, the moment of a successful strike, just after a gutter ball, and so on. Their goal: to identify what triggered the strongest emotional reactions, both positive and negative.

The M.I.N.D. Lab team expected that gamers would exhibit the stron-gest positive emotion when they earned high scores or when they completed difficult levels—in other words, during the triumphant fiero moments. The players did indeed show peaks of excitement and satisfaction during these moments. But the researchers noticed another set of positive emotion peaks that caught them off guard. They found that players exhibited the most potent combination of positive emotions when they made a mistake and sent the monkey ball veering off the side of the lane. Excitement, joy, and interest shot through the roof the second they lost their monkey ball.

Initially, the researchers were perplexed by the gamers' positive emotional reaction to "complete and unquestionable failure in the game." When we fail in real life, we are typically disappointed, not energized. We experience diminished interest and motivation. And if we fail again and again, we get more stressed, not less. But in *Super Monkey Ball 2*, failure seems to be more emotionally rewarding than success.

What was so interesting about failure in *Super Monkey Ball 2*? And why would it make gamers happier than winning?

The M.I.N.D. Lab interviewed the players and consulted with game designers in order to make sense of their findings. After much consideration, they concluded: failure in *Super Monkey Ball 2*, in an odd way, was something players could be proud of.

Whenever a player made a mistake in *Super Monkey Ball 2*, something very interesting happened, and it happened immediately: the monkey went whirling and wailing over the edge and off into space. This animation sequence played a crucial role in making failure enjoyable. The flying monkey was a reward: it made players laugh. But more importantly, it was a vivid demonstration of the players' agency in the game. The players hadn't failed passively. They had failed spectacularly, and entertainingly.

The combination of positive feeling and a stronger sense of agency made the players eager to try again. If they could send a monkey into outer space, then surely they could knock over a few bowling pins or roll over a few more bananas next time.

When we're reminded of our own agency in such a positive way, it's almost impossible not to feel optimistic. And that's the positive effect the researchers were measuring in the M.I.N.D. Lab: excitement, joy, and interest. The more we fail, the more eager we are to do better. The researchers were able to demonstrate this: the right kind of failure feedback is a reward. It makes us more engaged and more optimistic about our odds of success.

Positive failure feedback reinforces our sense of control over the game's outcome. And a feeling of control in a goal-oriented environment can create a powerful drive to succeed. Another player describes this phenomenon perfectly: "*Super Monkey Ball* is pretty much the dictionary definition of addictive. It brilliantly balances the intense frustration at failing to complete a course with the absolute desire to have 'just one more go.'"[4] To optimists, setbacks are energizing—and the more energized we get, the more fervently we believe that success is just around the corner. Which is why, on the whole, gamers just don't give up.

We aren't used to feeling so optimistic in the face of things that are extremely difficult for us. That's why so many gamers relish wickedly hard game content. Nearly every review you'll find of the *Super Monkey Ball* games praises them with descriptions such as "insanely frustrating" and "fiendishly difficult." We like it that way, precisely because it's so rare in real life to feel sincere, unabashed hope in the face of such daunting challenges.

It helps, of course, that gamers believe they have every chance of success when they sit down to play a new game. Justifiable optimism is built right in to the medium. By design, every computer and video game puzzle is meant to be solvable, every mission accomplishable, and every level passable by a gamer with enough time and motivation.

But without positive failure feedback, this belief is easily undermined. If failure feels random or passive, we lose our sense of agency—and optimism goes down the drain. As technology journalist Clive Thompson reminds us, "It's only fun to fail if the game is fair—and you had every chance of success."[5]

That's why Nicole Lazzaro spends so much time consulting with game developers about how, exactly, to design failure sequences that are spectacular and engaging. The trick is simple, but the effect is powerful: you have to show players their own power in the game world, and if possible elicit a smile or a laugh. As long as our failure is interesting, we will keep trying—and remain hopeful that we will succeed eventually.

Which gives us our next fix for reality:

⟳ FIX #4: BETTER HOPE OF SUCCESS

Compared with games, reality is hopeless. Games eliminate our fear of failure and improve our chances for success.

In many cases, that *hope* of success is more exciting than success itself.

Success is pleasurable, but it leaves us at a loss for something interesting to do. If we fail, and if we can try again, then we still have a mission.

Winning tends to end the fun. But failure? It keeps the fun going.

"Games don't last forever," says Raph Koster, a leading creative director of online games and virtual worlds. "I play something I'm good at, I get really far and do really well, then I get bored."[6] And that's when he stops playing and moves on to the next game. Why? Because being really good at something is less fun than being *not quite good enough—yet*.

Koster has written a book much beloved in the game industry, A *Theory of Fun for Game Design,* in which he argues that games are "fun" only as long as we haven't mastered them. He writes, "Fun from games arises out of mastery. It arises out of comprehension. . . . With games, *learning* is the drug." And that's why fun in games lasts only as long as we're not consistently successful.[7]

It's something of a paradox. Games are designed for us to learn them, get better at them, and eventually be successful. Any gamer who puts in the effort can't help but get better. And yet the better we are at a game, the less of a challenge it presents. Harder levels and tougher challenges can keep the feeling of "hard fun" alive for a while. But if we keep playing, we keep getting better—and so it's inevitable: the unnecessary obstacle becomes less of an obstacle over time.

That's why, Koster says, "the destiny of games is to become boring, not to be fun. Those of us who want games to be fun are fighting a losing battle against the human brain."[8] Fun will always morph into boredom, once we pass the critical point of being reliably successful. This is what makes games *consumable*: players wring all the learning (and fun) out of them.

Fun failure is a way to prolong the game experience and stretch out the learning process. Meanwhile, when we can enjoy our own failure, we can spend more time suspended in a state of urgent optimism—the moment of hope just before our success is real, when we feel inspired to try our hardest and do our best.

Learning to stay urgently optimistic in the face of failure is an important emotional strength that we can learn in games and apply in our real lives. When we're energized by failure, we develop emotional stamina. And emotional stamina makes it possible for us to hang in longer, to do much harder work, and to tackle more complex challenges. We need this kind of optimism in order to thrive as human beings.

Scientists have found that optimism is closely correlated to a higher quality of life in pretty much every way imaginable: better health, a longer life, less stress and anxiety, more successful careers, better relationships, more creativity, and more resilience in the face of adversity. This isn't surprising: optimism is what allows us to take action to improve our lives, and the lives of others. It allows us to flourish—to create the best life possible for ourselves. Flourishing isn't about pleasure or satisfaction; it's about living up to our fullest potential. And to truly flourish, we have to be optimistic about our own abilities and opportunities for success.

In fact, optimism about our own abilities not only makes us happier in the moment, it also increases our likelihood of success and feeling happier in the future. Numerous studies have shown that students, executives, and athletes are consistently more successful if they agree with statements like "I have the ability to change things with my actions" or "I am in control of my own fate."[9] Other studies show that when we're in an optimistic state of mind, we pay more attention, think more clearly, and learn faster.[10] Hope primes our minds for real success.

Of course, it's possible to go overboard: too much optimism can be as harmful as too little. We have to have the right level of optimism for the occasion. Martin Seligman recommends adopting what he calls "flexible optimism": continually assessing our abilities to achieve a goal, and intensifying or reducing our efforts accordingly.

When we practice flexible optimism, we see more opportunities for success—but we don't overstate our abilities, and we don't overestimate the amount of control we have over an outcome. And we reduce our optimism when we get feedback that we're pursuing unattainable goals or operating in a low-control environment. We recognize that our time and energy would be better spent elsewhere.

Games are perfect environments for practicing flexible optimism. And we *do* need help practicing more flexible optimism in everyday life. Randolph Nesse, a professor of evolutionary medicine at the University of Michigan, believes that our happiness depends on it—and has depended on it since the earliest days of human civilization.

Nesse's research focuses on the evolutionary origins of depression. Why does depression exist at all? If it's stayed in our gene pool for so long, he argues, there must be some evolutionary benefit. Nesse believes that depression may be an adaptive mechanism meant to prevent us from falling victim to blind optimism—and squandering resources on the wrong goals.[11] It's to our evolutionary advantage not to waste time and energy on goals we can't realistically achieve. And so when we have no clear way to make productive progress, our neurological systems default to a state of low energy and motivation.

During this period of mild depression, Nesse theorizes, we can conserve our resources and search for new, more realistic goals. But if we persist in pursuing unattainable goals? Then, Nesse proposes, the mechanism kicks into overdrive, triggering *severe* depression.

Nesse thinks this mechanism, and our tendency to set unrealistic goals, may be the cause of much of the current depression epidemic in the United States. We set extreme goals: fame, fortune, glory, and supersized personal achievements. We're encouraged, says Nesse, to believe that we can do anything we set our hearts to, and then we try to achieve dreams that are just unrealistic. We don't pay attention to our real skills and abilities, nor do we put our efforts toward the goals we are capable of achieving. We're distracted by extreme dreams—even when our evolutionary mechanism kicks in, signaling our ill-fated efforts.

But games can take us out of this depressive loop. They give us a *good* reason to be optimistic, satisfying our evolutionary imperative to focus on attainable goals. As happiness researcher Sonja Lyubomirsky writes, "We obtain maximum happiness when we take on flexible and appropriate goals."[12] Good games provide a steady flow of actionable goals in environments we know are designed for our success—and they give us the chance to inject some flexible and appropriate goals into our daily lives whenever we need them most.

The success we achieve in games is not, of course, real-world success. But for many people it is more *realistic* than the kinds of success we put pressure on ourselves to achieve—whether it's money, beauty, or fame.

It's depressing to spend our lives pursuing unrealistic goals. For anyone who wants to opt out of this culture of extreme dreaming, games help enormously: they shift our attention away from depressing goals and train us to be more flexibly optimistic. Today's best games help us *realistically* believe in our chances for success.

Of course, this might not be a perfect solution to the problem of unattainable goal setting in contemporary society. But in the meantime, it *does* make us feel better and builds our capacity for flexible optimism. We can opt out of whatever "the dream" is supposed to be, and focus our efforts instead on goals that give us real practice at working hard, getting better, and mastering something new.

Take, for example, the wildly popular video game *Rock Band 2*. The musical rhythm video game series has probably given us more exciting, realistic goals than any other video game in the history of the medium. It racked up more than a billion dollars of sales in its first year.[13] And along the way to becoming the number one best-selling game of 2009, as well as one of the most successful video games of all time, it has turned millions of players into aspiring hopefuls—*and* spectacular failures.

The Hope of Rock Star Success

To be a rock star is shorthand in our culture for supersized success. It's one of our favorite symbols of status and fame—and it's something that virtually none of us has any *real* hope of achieving. But when you play *Rock Band 2*, you get to aspire to rock stardom, with a knowing wink.

Rock Band is a game for up to four friends who perform the role of rock stars by singing into a microphone, banging on a plastic drum kit, and pressing out chords on plastic guitars with buttons instead of strings. You follow musical cues from the game, which tell you which combination of notes to hit — or sing — and when. All the while, your customizable rock star avatar appears on the screen, rocking out on a stage.

Of course, you won't become a real rock legend by playing a video game. But you do get to perform alongside friends and family, playing the hits and adopting the role of ultimate rock stars: the Who, the Grateful Dead, Pearl Jam, Nirvana, Guns N' Roses, the Rolling Stones, Bob Dylan, and the Beatles, to name a few. And you're performing the role in a way that feels genuine because it's such an active, hands-on experience.

When you are on vocals, you're singing the actual lyrics, and you have to sing well — the microphone has a pitch register to detect if you're singing the right notes at the right times. The drums are also a fairly realistic stand-in, with four different pressure-sensitive drum pads arranged at realistic heights, a drum pedal, and two cymbal add-ons. When you're hitting away with your drumsticks, trying to match the timing and arrangement of the falling notes on the screen, you really are making a rhythmic noise. Even professional drummers have remarked that it serves as a reasonable approximation of real rock drumming.[14]

As for the guitar and the bass — well, that's a more abstract kind of music making. Jesper Juul, a well-known academic researcher of video games, calls it the difference between "playing music," or actually making the musical sounds, and "performing music," or doing complex actions that get translated into musical sounds.[15]

Holding down different combinations of button frets *sort of* feels like using different frets to play different chords, and toggling the strum bar back and forth *sort of* feels like strumming the strings. More realistically, you're playing according to the same rhythms as a real guitarist would, and your fingers *are* moving deftly around the way they might on a real fret board. It's not playing real music, but it's performing musical and rhythmic actions in a way that feels closely connected to the actual song.

On all the instruments, the more accurately you hit your notes, the better and fuller the backing song tracks sound. If you're messing up on drums, the drum track disappears from the song. If you're hitting only half your guitar notes, the guitar track sounds spotty. But if the entire band is playing successfully, the song is virtually identical to the original artist's track.

Alex Rigopulos, one of the creators of the *Rock Band* series, has said that one of the goals of the game is for "the music to come alive when you're playing."[16] And that's the best way to describe the sensation of playing the game. Although you're not really "playing" the music, you *are* making the music come to life. You can really hear the impact of your efforts in the song that's produced—and hopefully you're making it sound better. The more complicated your finger work or rhythm work, or the more demanding the pitch detector is of you, the more real the connection between your work and how the song feels.

Every step up the difficulty chart gives you a more complicated set of musical actions to perform. And each added layer of complexity feels more closely connected to the real musical work of the song: more complex chords or deftly syncopated drumbeats or pitch perfection.

For all of these reasons, getting better at *Rock Band* feels like a truly worthy goal. You are mastering your favorite songs in a way that will let you connect and interact with them, *and* potentially perform them in front of an audience. This is perhaps the game's biggest secret to cultivating the hope of success: the *Rock Band* gaming culture extends many, many opportunities to perform in front of audiences. You can perform in front of friends and family in your own home. You can go to bars in just about every major city to participate in *Rock Band* game nights and perform on a real stage. And *Rock Band* is also one of the most popular choices for live video game tournaments these days.

The possibility of not only mastering a song, but also showing off that mastery to others, amplifies the optimistic drive. And fortunately for players who want to really master the hard and extreme versions of their favorite songs, failure in *Rock Band* is about as entertaining and energizing as failure can get.

The fun of *Rock Band* failure starts with the audio effects. If you get behind the beat or off your pitch or hit too many wrong notes, the song audibly starts to fall apart. First you hear bad notes in the musical accompaniment. Then you hear heckles and boos from the crowd. The more you fail, the more the song falls apart. Eventually, if you're bad enough, the visual effects kick in: you'll be booed off the stage by the animated crowd while your avatar pouts and skulks around the stage and all the band members shake their fists and scowl at each other. It's a highly entertaining fail sequence—so over-the-top that you can't help but laugh at yourself.

Even better, the "fail sequence" in real life is often more entertaining than the online version. When your performance power bar hits the red zone, indicating that you are about to be booed offstage, you can't help yourself: you give it one last shot, wailing away at your instrument the best you can. You fling the drumsticks around like crazy, or strain your vocal chords to the farthest extremes of your range, or mash the plastic guitar frets and thrash your strum button like a maniac. That last desperate rush of play is energizing even if you're playing alone—but when you're playing with at least one other person, it's also hysterically funny.

Combined with the positive emotions that this kind of spectacular failure sparks, you also get a bit of crucial information every time you wipe out on a song: an exact percentage readout of how far in the song you got before being booed off the stage. This information shows you what you've positively accomplished—even if it's only 33 percent, you've survived a third of the way through the song. You haven't so much failed as achieved partial success. And the higher your percentage gets, the more capable and confident you feel—and the desire for one more go at the song kicks in.

This desire isn't misdirected: the game environment supports your hope of success in several key ways. For instance, each player in your band can select a different difficulty level, meaning your experienced drummer can play on hard while your vocalist scrapes by on medium and your beginner guitarist takes it easy. This allows each player to set his or her own customized and realistic goal, while still playing the same song with the group.

If you're playing with friends, you can also "save" each other from musical disaster. If the bassist wipes out, the guitarist can play a spectacular solo and revive her. If the drummer fails, the singer can belt the chorus perfectly to bring him back into the game. You can actually fail and be revived by other band members twice before you get booed off for good. And, crucially, saves are dependent on the successful effort of another band member. One member's failure pushes the others to do better.

The real-time feedback from the game also makes it easy to learn from your mistakes. When you're playing the drum or string instruments, you get visual confirmation of every note you hit or miss. You know instantly that you're off the beat or getting your chords wrong, and it's easy to identify exactly where you're getting tripped up. When you're the singer, you can watch the musical staff to see if you're perfectly on pitch, sharp, or flat. You can adjust in real time, sliding up or down to hit the right note—and after a few goes at the same song, you actually start to sound better.

All of these features make *Rock Band* feel like a learning environment. And playing really does seem to help us grow musically, even if it's just a better understanding of the rhythms and tracks of songs we've always loved, or a greater confidence performing in front of friends and family.

Moreover, research suggests that players of music video games are increasingly driven to play *real* musical instruments. In a 2008 study of more than seven thousand *Rock Band* and *Guitar Hero* players, 67 percent of the nonmusicians in the group reported that they had been inspired to pick up a real instrument since they'd started playing the video games. Meanwhile, 72 percent of the gamers who considered themselves musicians reported that they'd spent more time playing their real instruments since beginning to play music video games.[17]

No major research has been published yet on whether games like *Rock Band* confer real musical ability. But these games are without a doubt conferring real optimism, which in turn inspires real musical participation.

YOU CAN PLAY *Rock Band* alone—practicing any of the four instruments by yourself—but gamer surveys indicate that hardly anyone does. In fact, a 2008 study by the Pew Internet & American Life Project looked at the role of video games in family life and explicitly credited music video games like *Rock Band* for increasing the amount of time gamers of all ages spend playing together.

It's no coincidence that one of the most optimism-building video games is also one of the most social. One of the biggest hopes we have for our own success is to share it. We want others to see our strengths, and to reflect our achievements back to us. Success, as they say, means nothing alone. For all the positive feedback that a game can give us, we crave the praise and admiration of our friends and family even more.

In fact, studies have shown that optimism makes us more likely to seek out social support and develop strong relationships.[18] When we feel a strong sense of agency and motivation, we draw other people closer into our lives. And that's why so much of the fun failure we experience in games is increasingly taking place in a social context. More and more, we are inviting our friends and family to play with us, whether it's in person or online. We seek out opportunities to perform our favorite games in front of audiences. And we form long-term teams, like our *World of Warcraft* guilds and our rock bands.

It may have once been true that computer games encouraged us to interact more with machines than with each other. But if you still think of gamers as loners, then you're not playing games.

CHAPTER FIVE

Stronger Social Connectivity

More than 5 million people are playing the online word game Lexulous on Facebook. And most of them are playing it with their moms. When the game was released in 2007, it became the first Facebook application to achieve a mass audience, and the familiarity of the gameplay was one of its main attractions. If you know how to play Scrabble, then you already know how to play Lexulous — it's just a slightly modified and unauthorized version of the classic board game, combined with online chat.[1] There's no time limit on turns, and games stay active even when you log out of the social network. Whenever it's your turn, Facebook sends you an alert to your home page, your e-mail, or your mobile phone.

Here's how one Lexulous reviewer sums up its cross-generational appeal: "Everyone in your social network, even your mom, knows how to play Scrabble."[2] No doubt that's why so many of the online rave reviews include the phrase "my mom" — like this one: "I live in Atlanta, and my mom's in Texas. We love to have game night across the miles. Although I am sure she needs a break from me kicking her butt all of the time. (Love you, Mom!)"[3]

I've been reading game reviews for most of my life and I've never seen anything close to this many mom references. In fact, it's not that much of a stretch to say that, for many, the primary reason they play *Lexulous* is to have an excuse to talk to their mom every day.

It's not just online reviews that have given me this suspicion—there's photographic evidence as well. Lexulous games are private, but players often post screenshots of their most triumphant moments on photo-sharing sites like Flickr and Photobucket. In these screenshots, which usually have titles like "Online Scrabble with Mom" or "In Which I Beat My Mother at Lexulous," you get a glimpse of the kind of everyday familial checking-in that runs alongside the wordplay.[4] Much of the chat is mundane game talk, but you also see a constant stream of catching up, like these messages spied on Flickr: "Have you started your internship yet? How is that going?"[5] and "Knee still hurt. Putting a lot of ice on it."[6] Or "What are you doing after work?" and "Your stepfather says hello."[7] Some chat messages simply express users' happiness to be playing together, like this one from a mom to two daughters: "Glad to see you two, even if you do spank me when we play. :)"[8] Of course, there are tons of messages that simply say: "I love you."[9]

Judging from the shared screenshots, it's not *just* moms whom players use Lexulous to keep in touch with daily. There are also plenty of running games against dads, cousins, siblings, in-laws, former coworkers, faraway friends, and spouses on business trips. (That's when I most frequently play Lexulous—I keep a game running against my husband when I'm traveling for work. It helps me feel like we're actually doing something together, not just checking in.)

Because you don't have to be online playing at the same time, it's easy to organize a game with anyone else, no matter where or how busy they are. You can easily keep up with the game by playing literally only a few minutes a day. And by keeping running games going with your real-life friends and family, you're ensuring daily opportunities to actively connect with the people you care about most.

The tight-knit nature of the Lexulous game world wasn't a necessary outcome of the game's design. On Facebook, you can technically start a Lexulous game with anyone—even people you don't know—but most people play against people they already count as Facebook "friends." Playing Lexulous is checking in with our loved ones, but with a purpose. For anyone who has ever needed a gentle reminder to stay in touch, Lexulous provides a motivation. It helps us stay actively connected, by reminding us that it's literally "our turn" to say something. And when there's a game on the line, suddenly staying in touch is not just pleasant and gratifying—it is also addictive.

The secret to the addictiveness of Lexulous is its *asynchronous* gameplay: players don't have to be online at the same time, and can take their turns whenever they want. Some Lexulous games go quickly, with players trading words every few minutes, but many games go quite slowly, with players taking just one or two turns a day, or even less often than that.

The unpredictable rhythm of asynchronous play adds a measure of anticipation. You're thinking about your next play, but you don't know when you'll be able to make it. You're motivated to act, but you have to wait for your Facebook friends to check back into the game. And because you often have no idea if your friends are still logged on or paying attention to the game, there's an emotional buildup to waiting for their next moves. As one player puts it, "You have to be addicted AND patient."[10]

The addictiveness of the game pushes us to initiate social interaction with members of our extended social network whom we might ordinarily leave out of our daily life online. Indeed, starting a new game with someone is making a commitment to interact with them at least a dozen or so times in the near future. And when you've got five or ten or twenty games going at once, you've effectively scheduled hundreds of microinteractions with people you like into your everyday routine.

According to user metrics reported in an article in the *Wall Street Journal*, on average one-third of registered Lexulous players at any given time have logged in at least thirty straight days in a row.[11] This is a measure of the remarkable *stickiness* of social network gaming—it capitalizes brilliantly on the increased motivation we feel when we play a good game. It leverages our increased interest and optimism to help us satisfy our often otherwise thwarted desire to feel more connected with friends and family.

Simply put, social network games make it both *easier* and *more fun* to maintain strong, active connections with people we care about but who we don't see or speak to enough in our daily lives.

Eric Weiner, an independent foreign correspondent and author of *The Geography of Bliss*, has covered happiness trends throughout the world. His research has confirmed for him that "our happiness is completely and utterly intertwined with other people: family and friends and neighbors. . . . Happiness is not a noun or verb. It's a conjunction. Connective tissue."[12] Games like Lexulous are intentionally designed to strengthen the connective tissue within our social networks. Each move we make in the game is a conjunction.

We clearly need more social conjunctions in our lives. As numerous economists and positive psychologists have observed, globally we make the mistake of becoming less social the richer we become as individuals, and as a society. As Weiner observes: "The greatest source of happiness is other people—and what does money do? It isolates us from other people. It enables us to build walls, literal and figurative, around ourselves. We move from a teeming college dorm to an apartment to a house and, if we're really wealthy, to an estate. We think we're moving up, but really we're walling off ourselves."[13]

Games like Lexulous can help us start chipping away at those walls. Lexulous was the first breakthrough social network game, but since its success, the genre has experienced dramatic growth—particularly on Facebook. In early 2010, a virtual farming game called FarmVille hit an astonishing benchmark: 90 million active players on Facebook, nearly 30 million of whom log in on any given day to harvest their virtual crops and tend to their virtual livestock.[14]

It's an unprecedented scale of participation in a single online game. Roughly one in seventy-five people on the planet is currently playing FarmVille, and one in two hundred people on the planet logs in on any given day to manage and grow their virtual farm. What accounts for this global popularity? FarmVille is the first game to combine the blissful productivity of *World of Warcraft* with the easy gameplay and social connectivity of Lexulous.

Half the fun of FarmVille is earning experience points and gold in order to level up and earn access to better crops and farm equipment, more exotic animals, and a bigger land plot. Every time you log in to the game, you can improve your stats by undertaking a series of simple, point-and-click tasks:

plow the soil, buy and plant the seeds, harvest the crops, pet your farm animals. Each crop takes between twelve hours and four days in real time to yield a harvest, so checking in every day or so becomes a regular habit. You start the game able to harvest just strawberries and soybeans on a humble two-by-six-square plot. Over time, you can work your way up to a "mighty plantation" plot of twenty-two by twenty-two squares, on which you can grow lilies, yellow melons, and coffee — not to mention care for bunny rabbits, pinto horses, and golden chickens.

But the real genius of FarmVille is the social layer on top of this immensely satisfying self-improvement work. The first time you log in to the game, you see a list of your real-life Facebook friends who are already tending their own virtual farms. You can make any or all of them your "neighbors" in the game and visit their farms whenever you want to see how they're doing.

You don't interact directly with these neighbors — instead, like most Lexulous play, FarmVille is an entirely asynchronous experience. While you're tending your own farm, pop-up windows nudge you to pay attention to your friends' and families' farms: "Chelsea could use help on her farm. Can you give her a hand?" or "Ralph's crops are looking a little puny. Could you please fertilize them?" Most players spend up to half their time in FarmVille helping others: raking up their leaves, shooing away raccoons, or feeding their chickens. You can also send your neighbors one free gift every day — a virtual avocado tree, a bale of hot pink hay, or a duck, for instance. Meanwhile, whenever you log back in to the game, you'll see a list of neighbors who have helped your farm, and you're likely to find a pile of presents to accept.

The gifts aren't real, of course. The favors don't help you in your everyday life. But the gesture isn't an empty one. Every gift or favor someone bestows upon you helps you achieve your goals in the game. And it's a virtuous circle. Every time you see that someone has helped your farm, you feel the urge to reciprocate. Over time, you build up a rhythm of checking in and helping others in your social network every single day.

It's not a good substitute for real interaction, but it helps keep extended friends and family in our daily lives when we might otherwise be too busy to stay connected. Games like Lexulous and FarmVille ensure we'll show up and do our part to nurture our relationships daily, and make a gesture of friendship whenever it's our turn.

And so we have our fifth fix for reality:

 FIX #5: STRONGER SOCIAL CONNECTIVITY

> Compared with games, reality is disconnected. Games build stronger social bonds and lead to more active social networks. The more time we spend interacting within our social networks, the more likely we are to generate a subset of positive emotions known as "prosocial emotions."

Prosocial emotions—including love, compassion, admiration, and devotion—are feel-good emotions that are directed toward others. They're crucial to our long-term happiness because they help create lasting social bonds.

Most of the prosocial emotions that we get from gaming today aren't necessarily built in to the game design; they're more of a side effect of spending more time playing together. Case in point: my husband and I first fell in love when we spent six weeks in each other's apartments playing a mystery adventure game called *Grim Fandango* on my laptop. Falling in love wasn't so much anything about that game in particular as it was a result of spending so much time working together to solve puzzles—not to mention negotiating who got to control the mouse and keyboard, and when—in order to lead us through the virtual world. Similarly, any pair or group of people who consistently play a game together, online or face-to-face, will have increased opportunities to express admiration for each other, to devote themselves to a common goal, to express sympathy for others' losses, and even to fall in love. (Which reminds me of the most interesting comment I've eavesdropped on by browsing Lexulous screenshots: "Quite a close game again. Loser has to marry the winner?"[15])

But beyond this kind of all-purpose social benefit to playing games together, there are two specific prosocial emotions that games give us: **happy embarrassment** and **vicarious pride**. Let's take a look at why these two prosocial emotions matter, and how online games generate them better than real-world interaction.

Happy Embarrassment

If there's one thing Lexulous players do even better than making obscure words out of random letters, it's gently teasing each other in a way that makes them feel good. And the most effective way they tease each other is through trash-talking.

Trash-talking, when it's a playful way to insult your competition, is almost as important to our enjoyment of social network games as the actual core gameplay. We crave the distinctly rewarding feeling we get from a good game when we soundly beat, or are beaten, by people we really like. More importantly, we crave the experience of teasing each other about it, in private *and* in public.

Consider, for example, the following public status updates from Lexulous players. These statements are visible to all members of their social network (including, no doubt, the people they are playing against), and sometimes to the whole world (which is how I happened to see them):

> "Playing Lexulous on Facebook with my mom. I'm winning. Hee hee hee!"[16]

> "I so pwnd my mom!"[17]

If you've never pwned your mom, you're clearly missing out.

To *pwn* someone—pronounced "pone" or "pawn," though most people just type it—means to achieve such a major victory you can't help but gloat afterward. It originates from a common typo of the word "own," since the letters *p* and *o* are next to each other on a standard keyboard; "own" has long been a popular gamer shorthand for the boastful comment "I'm so good at this game, I *own* it."[18]

Why is game pwning such an increasingly popular form of social interaction? And why, when we're on the receiving end, do we happily put up with it?

Teasing each other, recent scientific research has shown, is one of the fastest and most effective ways to intensify our positive feelings for each other. Dacher Keltner, a leading researcher of prosocial emotions at the University of California, has conducted experiments on the psychological benefits of teasing, and he believes that teasing plays an invaluable role in helping us form and maintain positive relationships.[19]

"The tease is like a social vaccine," Keltner explains. "It stimulates the recipient's emotional system." Teasingly trash-talking allows us to provoke each other's negative emotions in a very mild way—we stimulate a very small amount of anger or hurt or embarrassment. This tiny provocation has two powerful effects. First, it confirms trust: the person doing the teasing is demonstrating the capacity to hurt, but simultaneously showing that the intention is not to hurt. Just like a dog might play-bite another dog to show that it wants to be friends, we bare our teeth to each other in order to remind each other that we could, but never really *would*, hurt each other. Conversely, by allowing someone else to tease us, we confirm our willingness to be in a vulnerable position. We are actively demonstrating our trust in the other person's regard for our emotional well-being.

By letting someone tease us, we're also helping them feel powerful. We're giving them a moment to enjoy higher status in our social relationship—and humans are intensely attuned to shifts in social status. By letting someone else experience higher status, we intensify their positive feelings for us. Why? Because we naturally like people more when they enhance our own social status.

This is the essence of happy embarrassment and, according to Keltner's research, we're hardwired to feel it. He has documented the physiological basis for this complicated social effect in studies of face-to-face playful teasing and trash-talking. According to Keltner's findings, the recipient of the tease

almost invariably showed signs of lowered status, followed by an effort at rec-
onciliation: gaze aversion, bowed head, nervous smile, hand touching the
face, and so on. All of this is followed by a fleeting smile, a microexpression
that indicates we actually enjoy being teased by people we trust. Meanwhile,
the more obvious the display of lowered status, the more the teasers reported
liking the teased afterward.

None of this is a conscious process, Keltner's research shows. We mostly
tease and let ourselves be teased because it feels good. But the reason *why*
it feels good is that it builds trust and makes us more likable. Most of us
might not realize exactly why it enhances our social connection, but we definitely feel the emotional net positive after a teasing exchange. And this emotional reward encourages us to practice and repeat the behavior.

With all the pwnage and trash-talking happening in our favorite social
networking games, it's clear that they are giving us a perfect and much needed
space to practice and perform the good tease. Competitive games in particular
give us an excuse to adopt playful postures of superiority, and to let our friends
and family get away with the same.

We can also lower our status to strengthen our relationships by acting silly.
This helps explain the appeal of the popular video game genre known as
"party games." A party game is a game that's meant to be played socially, face-
to-face, and is easy to pick up the first time you try. *Rock Band* is one of the
most popular party games, and performing like a rock star—not to mention
failing a set—in front of friends and family definitely qualifies as a status-
raising or potentially happy-embarrassing moment.

Or consider *WarioWare: Smooth Moves* for the Wii, a game that is even
more physical than *Rock Band*. (The Wii remote controller has an accelerometer that detects hand movements, as well as optical sensors to know where
you're pointing the device.) Like most party games for the Wii, to play it you
have to perform it. *Smooth Moves* consists of more than two hundred different
"microgames" that require you to do a silly physical movement quickly: flap
your arms like a bird's wing, mime twirling a hula hoop, shove virtual dentures
into a virtual grandma's mouth. You have five seconds to figure out what

you're supposed to do, based on the images on the screen. Trying to think and move that quickly usually results in flailing around, goofy-looking gestures, and occasionally falling over.

Promotion screenshot and gameplay image of *WarioWare: Smooth Moves.*
(Nintendo Corporation, 2007)

One reviewer reasonably asks: "Games this crazy shouldn't be this popular, should they?"[20] But they are hugely popular. *Smooth Moves* has sold more than 2 million copies. They are easy to learn and quick to deliver emotional rewards—if you're willing to pick your virtual nose by shoving your game controller up and down, you really do trust the people around you.

Vicarious Pride

In a recent major study of more than one thousand gamers, a little-known prosocial emotion called "naches" ranked number eight on the top ten list of emotions that gamers say they want to feel while playing their favorite games.

Naches, a Yiddish word for the bursting pride we feel when someone we've taught or mentored succeeds, ranked just below surprise and fiero.[21]

The term "naches" hasn't caught on in the gamer world the way "pwn" or "fiero" has. But players in the study frequently described a kind of vicarious pride from playing over someone else's shoulder, and giving advice and encouragement—especially on games they themselves had already mastered. The author of the study, Christopher Bateman, an expert in both cognitive psychology and game design, adopted the term "naches" to describe this phenomenon, reporting, "Players seem to really enjoy training their friends and family to play games, with a whopping 53.4 percent saying it enhances their enjoyment."[22]

It's no surprise that mentoring our friends and family in gameplay makes us happy and brings us closer together. Paul Ekman, a pioneering emotions researcher and an expert on the phenomenon of naches, explains that this particular emotion is also likely an evolved mechanism, designed to enhance group survival. The happiness we get from cheering on friends and family ensures our personal investment in other people's growth and achievements. It encourages us to contribute to someone else's success, and as a result we form networks of support from which everyone involved benefits.[23] And because naches is so strongly correlated with survival, Ekman says, we feel it intensely. We don't describe ourselves as "bursting with pride" over our own success, but we do for others; this language suggests that the feeling of naches is even more explosive than personal fiero.

However, we don't naturally explode with pride at someone else's success if we haven't helped and encouraged them; too often, we feel jealousy or resentment. If we aren't actively contributing to the achievement with our support, then our emotional systems don't register vicarious pride. To generate the emotional reward of naches, we have to throw ourselves into the act of mentoring.

Most parents live in an almost constant state of naches. Unfortunately, outside of parenthood, we aren't always alert to opportunities for naches—among friends, between husband and wife, or from children toward their parents—because we don't have significant incentive or encouragement to mentor each other in everyday school or work. For the most part, we live in a culture of individual achievement, or what Martin Seligman calls "the waxing of the self" and "the waning of the commons."[24] He explains, "The society we live in takes the pleasures and pains, the successes and failures of the individual with unprecedented seriousness."[25] And when we see success or failure as an entirely individual affair, we don't bother to invest time or resources in someone else's achievements.

We need more naches, which helps explain the rise in single-player games being played with two or more people in the same room. Game researchers who study industry trends report that, increasingly, one person will play a game while another, or others, watch, encourage, and advise.[26] What makes this scenario attractive—and here is a big difference between ordinary life and games—is that computer and video games are perfectly replicable obstacles, we know in advance that our support will be useful, and we know exactly what our friends and family members are getting themselves in for.

The notoriously difficult puzzle game *Braid*, by independent game developer Jonathan Blow, is a perfect example of this phenomenon. Players must work their way through thirty-seven monster-filled puzzle rooms in order to rescue a princess. Early reviews of the single-player game were raves, but many reviewers worried that the reliance on puzzles would limit the replay value of the game. Once you'd solved a puzzle, one reviewer wrote, "there is little incentive to come back for seconds."[27]

But a large amount of anecdotal evidence from gamer blogs and forums suggests that gamers are revisiting *Braid*—in order to generate naches. Players seem absolutely tickled to watch friends and family work out the a-ha moments for each puzzle, lending their advice and positive morale in the face of the game's frustrating mental challenges. "Just finished the game, now I'm watching my wife work through it and it's a delight," one husband-turned-mentor writes.[28] Another says, "I finished the game last night and only needed help from my kids on two of the very final puzzle pieces. I think they were very proud of their mom!"[29] Games give us the opportunity to learn and master new challenges, and usually we learn skills that we can pass on to the other gamers in our lives.

Not all the social rewards we get from playing games are about strengthening bonds with people we already know. Social contact with strangers can offer different kinds of emotional reward, at the right times. One of these rewards that is unique to massively multiplayer online game environments is something researchers call "ambient sociability." It's the experience of playing alone together, and it's a kind of social interaction that even the most introverted among us can enjoy.

Ambient Sociability

Sometimes we want company, but we don't want to actively interact with anybody. That's where the idea of playing alone together comes in.

MMOs are famous for their collaborative quests and group raids. But it turns out that a majority of players prefer to play the game solo. An eight-month study of more than 150,000 *World of Warcraft* players discovered that players were spending on average 70 percent of their time pursuing individual missions, barely interacting with other players.[30] The researchers, based at Stanford University and Palo Alto Research Center (PARC), found this surprising and counterintuitive. Why bother paying a monthly subscription to participate in a massively multiplayer game world if you are going to ignore the masses?

The researchers conducted interviews to explore these findings and found that players enjoyed *sharing* the virtual environment, even if there was little to no direct interaction. They were experiencing a high degree of "social presence," a communications theory term for the sensation of sharing the same space with other people.[31] Although the players were not fighting each other or questing together, they still considered each other virtual company. The Stanford and PARC research team dubbed this phenomenon "playing alone together."[32]

One *World of Warcraft* player explains on her blog why she prefers to play alone together: "It's the feeling of not being alone in the world. I love being around other real players in the game. I enjoy seeing what they're doing, what they've achieved, and running across them out in the world 'doing their thing' while I'm doing mine."[33] What she describes here is actually a special kind of social presence: a presence enhanced by sharing goals and engaging in the same activities. The players can *recognize* each other because they have a common understanding of what they're doing and why. Their actions are intelligible and meaningful to each other.

Ambient sociability is a very casual form of social interaction; it may not create direct bonds, but it does satisfy our craving to feel connected to others. It creates a kind of social expansiveness in our lives—a feeling of inclusion in a social scene, and access to other people if we want it. The Stanford and PARC researchers posited that introverted players were more likely to enjoy playing alone together, and recent cognitive science studies support this theory. The best explanation scientists have for why some people are extroverted while others are introverted has to do with two differences in brain activity.

First, introverts in general tend to be more sensitive to external sensory stimulus: the cortical region of the brain, which processes the external world of objects, spaces, and people, reacts strongly in the presence of any stimulus. Extroverts, on the other hand, have lower cortical arousal. They require *more* stimulus to feel engaged with the external world. This makes extroverts more likely to seek higher levels of social stimulation, while introverts are more likely to feel mentally exhausted after lower levels of social engagement.

Meanwhile, extroverts tend to produce more dopamine in response to social rewards—smiling faces, laughter, conversation, and touch, for example. Introverts, in turn, are less sensitive to these social reward systems but highly sensitive to *mental* activity, such as problem solving and puzzling and solo exploration. Researchers say this explains why extroverts seem happier around other people and in stimulating environments: they are feeling significantly more intense positive emotions than introverted people.

But some game researchers, including Nicole Lazzaro, believe that ambient sociability and lightweight social interaction can actually train the brain to experience social interaction as more rewarding. Lazzaro proposes that since introverts are so sensitive to the rewards of mental activity, which gaming provides, doing these activities in online social settings can create new, positive associations for introverts about social experience. In other words, games like WoW may make introverts feel more comfortable with social interaction in general.

Studies have yet to be conducted to offer concrete support to this theory, but initial interviews and anecdotal evidence suggest it is worth further investigation. Our solo WoW player describes how she can be drawn into lightweight social interaction even as she makes her own way in the online world: "Chuck a heal there, apply a buff here, kill that thing that's about to kill that player, ask for some quick help or information, join up for a spontaneous quick group."[34] She remains open to these unexpected social interactions, and they are an essential part of why she likes to play alone together. She craves the *possibility* of "the spontaneous adventures that erupt between real people."

Why does this matter? Why is it a good thing for introverts to be open to more social interaction, and to find shared experiences more rewarding?

In study after study, positive-psychology researchers have shown that extroversion is highly correlated with greater happiness and life satisfaction. Extroverts are simply more likely to seek out the experiences that create social bonding and affection. As a result, they are better liked and better supported than introverts, two measures that factor heavily into quality of life. Introverts want to be liked and appreciated, and they need help just as much as anyone else; they're just not as motivated to seek out opportunities to build up that kind of positive social feeling and exchange.[35]

Fortunately, as many gamers are discovering, ambient sociability can play a key role in building up a desire for social interaction in the most introverted of people. Ambient sociability is hardly a substitute for real-world social interaction. But it can serve as a gateway to real-world socializing—and therefore greater quality of life—by helping introverts learn to view social engagement as more intrinsically rewarding than they are naturally predisposed to do.

GAME DESIGNER Daniel Cookman writes that when gamers decide to play with strangers or with people they know in real life, they're effectively choosing between "forging new relationships or strengthening old ones. . . . We can ask which the stronger draw is: strong, safe relationships with existing friends, or weak, 'risky' relationships with new people." Cookman says that, in most circumstances, he (and most gamers) prefers to strengthen existing relationships. The payoff is simply greater, and more clearly connected to our everyday lives.

Cookman is right that, on the whole, gamers make the choice to strengthen existing relationships—increasingly, online gamers report that they prefer to play online with people they know in real life. This is truer the younger a gamer you are. A recent three-year study of Internet use by young people in the United States revealed that gamers under eighteen spend 61 percent of their game time playing with real-life friends and family, rather than alone or with strangers.[36]

But Cookman acknowledges that there is another factor to consider. Play with strangers or play with friends? "In order to answer this question in any meaningful fashion," he writes, "you first need to answer a more personal question. 'Are you lonely?'"[37]

We can't discuss the social rewards of gaming without mentioning the positive role they play in helping us combat our feelings of loneliness. As a general rule, we'd rather play with friends. But if that isn't possible, we'll take strangers any day over playing alone. Cookman sums up the prevailing sentiment: "I'm not sure if having a stranger yell at me in [a first-person shooter game] will result in any long-lasting friendships, but it is certainly better than being alone."[38]

The gamer website Pwn or Die, popular with an audience of teenagers and young adults, has a short manifesto on "Ways Video Games Actually Benefit 'Real Life.'" At the top of the list is simply staving off loneliness. "When there are no kids in the neighborhood, it is late at night, or your best friend is miles away, video games give you an opportunity to interact with other people and be social."[39]

Would it be more rewarding to have a real-world space in which to have face-to-face interaction? Probably—there is significant evidence to suggest that social rewards are intensified by things like eye contact and touch. But face-to-face contact isn't always possible. Moreover, if we're feeling depressed or lonely, we might not have the emotional reserves to get up and get out, or to contact a real-life friend or family member. Playing a game online, like ambient sociability, can be a stepping-stone to a more positive emotional state and, with it, more positive social experiences.

FIFTEEN YEARS AGO, political scientist Robert Putnam famously worried that the United States was turning into a nation of people who go "bowling alone." In his hugely influential book about the collapse of extended community, he documented a worrying trend: that we are increasingly likely to hunker down and prefer the company of just a few people rather than participate in civic organizations or in a larger social context in general.

Putnam considered the collapse of extended community in our everyday lives to be a major threat to our quality of life, and he made this point so persuasively that, for years since, experts have debated the best ways to reverse it. Public institutions have also tried everything possible to rebuild the traditional community infrastructure. But, as gamers are finding out, rebuilding traditional ways of connecting might not be the solution — reinvention might work better.

Gamers, without a doubt, are reinventing what we think of as our daily community infrastructure. They're experimenting with new ways to create social capital, and they're developing habits that provide more social bonding and connectivity than any bowling league ever could.

As a society, we may feel increasingly disconnected from family, friends, and neighbors — but, as gamers, we are adopting strategies to reverse the phenomenon. Games are increasingly a crucial social thread woven throughout our everyday lives. We're using asynchronous social interaction in games like Lexulous and FarmVille to build stronger, stickier social connections. We're spending more time teasing and mentoring each other in games like *Smooth Moves* and *Braid*, in order to build trust and intensify our social commitments. And we're creating worlds of ambient sociability, as in *World of Warcraft*, where even the most introverted among us have opportunities to develop their social stamina and get more social connectivity in their lives.

Gamers, emphatically, are *not* gaming alone.

And the more we game together, the more we get the sense that we're creating a global community with a purpose. Gamers aren't just trying to win games anymore. They have a bigger mission.

They're on a mission to be a part of something *epic*.

CHAPTER SIX

Becoming a Part of Something Bigger Than Ourselves

In April 2009, *Halo 3* players celebrated a collective spine-tingling milestone: 10 billion kills against their virtual enemy, the Covenant. That's roughly one and a half times the total number of every man, woman, and child on earth.

To reach this monumental milestone, *Halo 3* players spent 565 days fighting the third and final campaign in the fictional Great War, protecting earth from an alliance of malevolent aliens seeking to destroy the human race. Together, they averaged 17.5 million Covenant kills a day, 730,000 kills per hour, 12,000 kills a minute.

Along the way, they'd assembled the largest army on earth, virtual or otherwise. More than 15 million people had fought on behalf of the science fiction game's United Nations Space Command. That's roughly the total number of active personnel of all twenty-five of the largest armed forces in the real world, combined.[1]

Ten billion kills wasn't an incidental achievement, stumbled onto blindly by the gaming masses. *Halo* players made a concerted effort to get there. They embraced 10 billion kills as a symbol of just how much the *Halo* community could accomplish—and they wanted it to be something bigger than anything any other game community had achieved before. So they worked hard to make every single player as good at *Halo* 3 as possible. Players shared tips and strategies with each other and organized round-the-clock "co-op," or cooperative, campaign shifts. They called on every registered member of *Halo* online to pitch in: "This could be something big, but we will need YOU to get it done."[2] They treated their mission like an urgent duty. "We know we'll be doing our part," one game blog declared. "Will you?"[3]

It's no wonder London *Telegraph* reporter Sam Leith observed in his coverage of the *Halo* 3 community that "a big shift has taken place, in recent years, in the way video games are played. What was once generally a solitary activity is now . . . overwhelmingly a communal one."[4] More and more, gamers aren't just in it for themselves. They're in it for each other—and for the thrill of being a part of something bigger.

When *Halo* players finally reached their goal, they flooded online forums to congratulate each other and claim their contributions. "I just did some math and with my 32,388 kills I have .00032% of the 10 billion kills," one player wrote. "I feel like I could have contributed more . . . well, on to 100 billion then!"[5] This reaction was typical, and the new 100 billion goal was repeated widely on *Halo* forums. Fresh off one collective achievement, *Halo* players were ready to tackle an even more monumental goal. And they were fully prepared to recruit an even bigger community to do it. As one gamer proposed: "We did that with just a few million gamers. Imagine what we could do with the full force of six billion humans!!"[6]

Halo's creators, a Seattle, Washington–based game studio called Bungie, joined in the celebration. They issued a major press release and an open letter to the *Halo* community, emphasizing the teamwork it had taken to get to 10 billion kills: "We've hit the Covenant where it hurts. Made them pay a price for setting foot on our soil. We're glad we've got you by our side, soldier. Mighty fine work. Here's to ten billion more."[7]

Perhaps you're thinking to yourself right now: *So?* What's the point? The Covenant isn't real. It's just a game. What have the players actually *done* that's worth celebrating?

On one hand, nothing. There's no *value* to a Covenant kill, whether you score one, 10 billion, or even 100 billion of them. Value is a measure of importance and consequence. And even the most die-hard *Halo* fan knows that there's no real importance or consequence to saving the human race from a fictional alien invasion. There's no actual danger being averted. There are no real lives being saved.

But on the other hand, just because the kills don't have value doesn't mean they don't have *meaning*.

Meaning is the feeling that we're a part of something bigger than ourselves. It's the belief that our actions matter beyond our own individual lives. When something is meaningful, it has significance and worth not just to ourselves, or even to our closest friends and family, but to a much larger group: to a community, an organization, or even the entire human species.

Meaning is something we're all looking for more of: more ways to make a difference in the bigger picture, more chances to leave a lasting mark on the world, more moments of awe and wonder at the scale of the projects and communities we're a part of.

How do we get more meaning in our lives? It's actually quite simple. Philosophers, psychologists, and spiritual leaders agree: the single best way to add meaning to our lives is to *connect our daily actions to something bigger than ourselves*—and the bigger, the better. As Martin Seligman says, "The self is a very poor site for meaning." We can't *matter* outside of a large-scale social context. "The larger the entity you can attach yourself to," Seligman advises, "the more meaning you can derive."[8]

And that's exactly the point of working together in a game like *Halo 3*. It's not that the Covenant kills have value. It's that pursuing a massive goal alongside millions of other people feels good. It feels meaningful. When players dedicate themselves to a goal like 10 billion Covenant kills, they're attaching themselves to a cause, and they're making a significant contribution to it. As the popular gamer site Joystiq reported on the day *Halo* players celebrated their 10 billionth kill: "Now we know for sure. . . . Every kill you get in *Halo 3*'s campaign actually *means* something."[9]

To experience *real* meaning, we don't have to contribute something of *real* value. We just have to be given the opportunity to contribute at all. We need a way to connect with others who care about the same massively scaled goal we do, no matter how arbitrary the goal. And we need a chance to reflect on the truly epic scale of what we're doing together.

Which gives us our sixth fix for our broken reality:

 FIX #6: EPIC SCALE

> Compared with games, reality is trivial. Games make us a part
> of something bigger and give epic meaning to our actions.

"Epic" is the key word here. Blockbuster video games like *Halo*—the kind of games that have a production budget of thirty, forty or even fifty million dollars—aren't just "something bigger." They're big enough to be *epic*.

Epic is one of the most important concepts in gamer culture today. It's how players describe their most memorable, gratifying game experiences. As one game critic writes, "*Halo 3* is epic. It empowers you the way no other game can. It doesn't have moments, but events. Experiences that tickle the soul, sending shivers down the spine."[10]

A good working definition for "epic" is something that far surpasses the ordinary, especially in size, scale, and intensity. Something epic is of *heroic proportions*. Blockbuster video games do epic scale better than any other medium of our time, and they're epic in three key ways:

> They create *epic contexts for action*: collective stories that help us connect our individual gameplay to a much bigger mission.

> They immerse us in *epic environments*: vast, interactive spaces that provoke feelings of curiosity and wonder.

> And they engage us in *epic projects*: cooperative efforts carried out by players on massive scales, over months or even years.

There's a reason why gamers love epic games. It's not just that bigger is better. It's that bigger is more awe-inspiring.

Awe is a unique emotion. According to many positive psychologists, it's the single most overwhelming and gratifying positive emotion we can feel. In fact, neuropsychologist Paul Pearsall calls awe "the orgasm of positive emotions."[11]

Awe is what we feel when we recognize that we're in the presence of something bigger than ourselves. It's closely linked with feelings of spirituality, love, and gratitude—and more importantly, a desire to serve.

In *Born to Be Good*, Dacher Keltner explains, "The experience of awe is about finding your place in the larger scheme of things. It is about quieting the press of self-interest. It is about folding into social collectives. It is about feeling reverential toward participating in some expansive process that unites us all and that ennobles our life's endeavors."[12]

In other words, awe doesn't just *feel* good; it inspires us to *do* good.

Without a doubt, it's awe that a *Halo 3* player is feeling when he says that the game sends "shivers down the spine." Spine tingling is one of the classic physiological symptoms of awe—along with chills, goose bumps, and that choked-up feeling in the throat.

Our ability to feel awe in the form of chills, goose bumps, or choking up serves as a kind of emotional radar for detecting meaningful activity. Whenever we feel awe, we know we've found a potential source of meaning. We've discovered a real opportunity to be of service, to band together, to contribute to a larger cause.

In short, awe is a call to collective action.

So it's no accident that *Halo* players are so inclined toward collective efforts. It's the direct result of the game's epic, and awe-inspiring, aesthetic. Today's best game designers are experts at giving individuals the chance to be a part of something bigger—and no one is better at it than the creators of *Halo*. Everything about the *Halo* games—from the plot and the sound track to the marketing and the way the community is organized online—is intentionally crafted to make players feel that their gameplay really means something. And the one simple trick used over and over again is this: always connect the individual to something bigger.

Let's take a closer look at exactly how *Halo* does it.

Epic Context for Heroic Action

> It's five hundred years in the future. The Covenant, a hostile alliance of alien species, is hell-bent on destroying humanity. You are Master Chief Petty Officer John 117—once an ordinary person, now a supersoldier, augmented with biological technologies that give you superhuman speed, strength, intelligence, vision, and reflexes. Your job is to stop the Covenant and save the world.

That's the basic *Halo* story. It's not that different from many other blockbuster video games. As veteran game developer Trent Polack puts it, "To look at the majority of games today, one might think that gamers care only about saving the world." He would know: some of Polack's previous games have asked players to save the galaxy from malevolent aliens (*Galactic Civilizations II*), save the universe from evil deities (*Demigod*), and save the world from marauding Titans (*Elemental: War of Magic*).

Why *are* so many games about saving the world? In an industry article about the rise of "epic scale" narratives in video games, Polack suggests, "When games give players the epic scope of saving the galaxy, destroying some reawakened ancient evil, or any other classical portrayal of good versus evil on a grand scale, they're fulfilling gamers' power fantasies."[13]

I agree with Polack, but it's important that we be clear on exactly what *kind* of power fantasy is being fulfilled by these save-the-world stories.

Any video game that features a slew of high-powered weapons and gameplay that consists largely of shooting and blowing things up is, at one level, about the aesthetic pleasures of destruction and the positive feelings we get from exerting control over a situation.[14] This is true of any shooter game on the market today. But we don't need an epic story about saving the world to get those pleasures. We can get them quite effectively, and more efficiently, from a simple, plotless game like Atari's *Breakout*. Games that come with epic, save-the-world narratives are using them to help players get a taste of a different kind of power. It's the power to act with meaning: to do something that matters in a bigger picture. The story is the bigger picture; the player's actions are what matters.

As Polack explains, "Story sets the stage for meaning. It frames the player's actions. We, as designers, are not telling, we're not showing, we're informing the *doing*—the actions that players engage in and the feats they undergo." These feats make up the player's story, and the story is ultimately what has meaning.

Not every game feels like a larger cause. For a game to feel like a *cause*, two things need to happen. First, the game's story needs to become a **collective context** for action—shared by other players, not just an individual experience. That's why truly epic games are always attached to large, online player communities—hundreds of thousands or millions of players acting in the same context together, and talking to each other on forums and wikis about the actions they're taking. And second, the actions that players take inside the collective context need to feel like **service**: every effort by one player must ultimately benefit all the other players. In other words, every individual act of gameplay has to eventually add up to something bigger.

Halo is probably the best game in the world at turning a story into a collective context and making personal achievement feel like service.

Like many other blockbuster video games, *Halo* has extensive online community features: discussion forums, wikis, and file sharing (so that players can upload and share videos of their finest gameplay moments). But Bungie and Xbox have taken it much further than these traditional context-building tools. They've given players groundbreaking tools for tracking the magnitude of their collective effort and unprecedented opportunities to reflect on the epic scale of their collective service.

Every *Halo* player has their own story of making a difference, and it's documented online in their "personal service record." It's an exhaustive record and analysis of their individual contributions to the *Halo* community and to the Great War effort—or as Bungie calls it, "Your entire *Halo* career."

The service record is stored on the official Bungie website, and it's fully viewable by other players. It lists all the campaign levels you've completed, the medals you've earned, and the achievements you've unlocked. It also includes a minute-by-minute, play-by-play breakdown of *every single* Halo *level or match you've ever played online*. For many *Halo* players, that means thousands of games over the past six years—ever since the *Halo* series first went online in 2004—all laid out and perfectly documented in one place.

And it's more than just statistics. There are data visualizations of every possible kind: interactive charts, graphs, heat maps. They help you learn about your own strengths and weaknesses: where you make the most mistakes, and where you consistently score your biggest victories; which weapons you're most proficient with, and which you're weakest with; even which teammates help you play better, and which don't.

Thanks to Bungie's exhaustive data collection and sharing, everything you do in *Halo* adds up to something bigger: a multiyear history of your own personal service to the Great War.

But it's not just your history—it's much bigger than that. You're contributing to the Great War effort alongside millions of other players, who also have service records online. And *service* really is a crucial concept here. A personal service record isn't just a profile. It's a history of a player's contributions to a larger organization. The fact that your profile is called a "service record" is a constant reminder. When you play *Halo* online, rack up kills, and accomplish your missions, you're *contributing*. You're actively creating new moments in the history of the Great War.[15]

The moments all add up. The millions of individual personal service records taken together tell the real story of *Halo*, a collective history of the Great War. They connect all the individual gamers into a community, a network of people fighting for the same cause. And the unprecedented scale of data collected and shared in these service records underscores just how epic the players' collective story is. Bungie recently announced to players that its personal-service-record servers handled more than 1.4 *quadrillion* bytes of data requests from players in the past nine months. That's 1.4 petabytes in computer science terms.

To put that number in perspective, experts have estimated that the entire written works of humankind, from the beginning of recorded history, in all languages, adds up to about 50 petabytes of data.[16] *Halo* players aren't quite there yet—but it's not a bad start, considering that they've been playing together online for only six short years, compared to all of recorded human history.

One of the best examples of innovative collective context building is the *Halo* Museum of Humanity, an online museum that purports to be from the twenty-seventh century, dedicated to "all who fought bravely in the Great War." Of course, it's not a real museum; it was developed by the Xbox marketing group to build a more meaningful context for *Halo 3*.

The museum features a series of videos done in the classic style of Ken Burns' *Civil War* series: interviews with Great War veterans and historians, images from Covenant battles, all set to a hymnal score. As one blogger wrote, "The videos in the *Halo* Museum of Humanity seem like they could have been pulled straight from The History Channel. . . . It's nice to see video game lore treated with this kind of reverence."[17]

Reverence—the expression of profound awe, respect and love, or veneration— is usually an emotion we reserve for very big, very serious things. But that was precisely the point of the *Halo* Museum of Humanity: to acknowledge how seriously *Halo* players take their favorite game, and to inspire the kind of epic emotions that have always been the best part of playing it.

It's worked. The video series packs a real emotional wallop, despite the fact that, in the words of one player, "it's meant to honor heroes that never existed."[18] Brian Crecente, a leading games journalist, wrote, "It left me with chills."[19] And online forums and blogs were full of comments expressing heartfelt emotion. One player put it best when he wrote, "Really poignant. They've made something real out of fiction."[20]

It's not that the museum is such a believable artifact from the future. It's that the *emotions* it provokes are believable. The online Museum of Humanity is a place to reflect on the extreme scale of the *Halo* experience: the years of service, the millions of players involved. The Great War isn't real, but you really do feel awe when you think about the scale of the effort so many different people have made to fight it.

In the end, as one player sums it up, "*Halo* proves that you can have a shooter game with a story that really means something. It draws you in and makes you feel like you're part of something bigger."[21]

But *Halo* isn't just a bigger story. It's also a bigger environment—and this brings us to our next strategy for connecting players to something bigger: built epic environments, or highly immersive spaces that are intentionally designed to bring out the best in us.

Epic Environments—Or How to Build a Better Place

An epic environment is a space that, by virtue of its extreme scale, provokes a profound sense of awe and wonder.

There are plenty of natural epic environments in the world: Mount Everest, the Grand Canyon, Victoria Falls, the Great Barrier Reef, for example. These spaces humble us; they remind us of the power and grandeur of nature, and make us feel small by comparison.

A *built* epic environment is different: it's not the work of nature, but rather a feat of design and engineering. It's a *human* accomplishment. And that makes it both humbling and empowering at the same time. It makes us feel smaller as individuals, but it also makes us feel capable of much bigger things, together. That's because a built epic environment—like the Great Wall of China, the Taj Mahal, or Machu Picchu—is the result of extreme-scale collaboration. It's proof of the extraordinary scale of things humans can accomplish together.

Halo 3 is, without a doubt, such an environment.

The game consists of thirty-four different playing environments spanning more than two hundred thousand light-years of virtual space. From one level to the next, you might find yourself traveling from the crowded market city of

Voi, Kenya, to the Ark, a desert far, far beyond the limits of our own Milky Way galaxy.

It's not just how big the *Halo* playing field is; it's also how diverse and carefully rendered the environments are. As Sam Leith observes, "The building of a game like *Halo 3* is a work of electronic engineering comparable in scale to the building of a medieval cathedral." It took Bungie three years to craft this gaming cathedral, with a team of more than 250 artists, designers, writers, programmers, and engineers collaborating together. "You get a sense of the scale and intricacy of the task," Leith continues, "by considering the sound effects alone: The game contains 54,000 pieces of audio and 40,000 lines of dialogue. There are 2,700 different noises for footsteps alone, depending on whose foot is stepping on what."[22]

And that's what players are appreciating when they get goose bumps from *Halo*: the unprecedented achievement it represents as a work of computer design and engineering. Gamers aren't so much in awe of the environment itself as they are in awe of the work and dedication and vision required to create it. In this regard, *Halo* players join a long tradition in human culture of feeling awe, wonder, and gratitude toward the builders of epic environments.

THE VERY FIRST epic environments were constructed more than eleven thousand years ago, during the Neolithic period, or the New Stone Age. In other words, six thousand years before humans first used the written word, they were already building physical spaces to inspire awe and cooperation.

The world's oldest known example of an epic built environment is the Gobekli Tepe. Discovered less than two decades ago in southeastern Turkey, it's believed to predate Stonehenge by a staggering six thousand years. It's a twenty-five-acre arrangement of at least twenty stone circles, between ten and thirty meters in diameter each, made from monolithic pillars three meters high.

In comparison with other stone houses, tombs, and temples from the same period and location, this building was constructed on an extreme scale: it was much, *much* bigger, taller, and more formidable in its design than anything archaeologists had seen before at the time of its discovery. One archaeologist on the scene described it as "a place of worship on an unprecedented scale— humanity's first 'cathedral on a hill.'"[23]

And it wasn't just the scale of the building—it was its particular winding design. The Gobekli Tepe features an intricate series of passageways that would lead visitors through the dark to a cross-shaped inner sanctum, almost like a labyrinth. This particular architecture seems designed intentionally to trigger interest and curiosity, alongside a kind of trembling wonder. What would be around the next corner? Where would the path take them? They would need to hold on to other visitors for support, feeling their way through the darkness.

Crucially, the Gobekli Tepe wasn't an isolated example. As researchers have discovered since, epic stone cathedrals were common across the Neolithic landscape. Most recently, in August 2009, archaeologists working in northern Scotland unearthed the ruins of a 5,330-square-foot stone structure with twenty-foot ceilings and sixteen-foot-thick walls, also of a labyrinthine design, and also dating back to the New Stone Age.[24] "A building of this scale and complexity was here to amaze, to create a sense of awe in the people who saw this place," Nick Card, director of the archaeological dig, said to reporters when the ancient cathedral was first unearthed.

In the wake of unearthing these types of structures all over the planet, archaeologists have recently proposed a startling theory: that these stone cathedrals served an important purpose in the evolution of human civilization. They actually inspired and enabled human society to become dramatically more cooperative, completely reinventing civilization as it once existed. In an in-depth report in *Smithsonian* magazine on these Neolithic cathedrals, Andrew Curry wrote:

> Scholars have long believed that only after people learned to farm and live in settled communities did they have the time, organization and resources to construct temples and support complicated social structures. But . . . [perhaps] it was the other way around: the extensive, coordinated effort to build the monoliths literally laid the groundwork for the development of complex societies.[25]

In fact, as Curry quotes one scientist in his article, "You can make a good case these constructs are the real origin of complex Neolithic societies."[26]

No wonder epic environments inspire gamers today to collective efforts. They have been inspiring humans to work together to do amazing things for eleven thousand years and counting.

SO VIDEO GAMES didn't invent epic environment design. They inherited the tradition from some of our earliest ancestors. But they *are* making epic environments remarkably more accessible, to vastly more people, on a daily basis.

Archaeologists say that worshippers would have traveled more than a hundred miles by foot to visit the Gobekli Tepe, and they may have visited it just once in a lifetime. Today, however, it's easy to immerse ourselves in epic environments whenever we want. Instead of traveling great distances for a single encounter with a physical cathedral, we can instantly transport ourselves there from anywhere in the world, simply by loading up a blockbuster video game.

Our experience of these epic game environments isn't physical, but it is real in one crucial sense. The engineering of the virtual environment represents, today, a collaborative feat on an extreme scale. It takes an extraordinary collective and coordinated effort to create these virtual worlds—years of full-time, painstaking work by hundreds of artists and programmers—and the first time a gamer enters one of these massive environments, they are experiencing real awe at the ability of ordinary people, when they band together, to create extraordinary spaces.

Meanwhile, video game developers have evolved the art of epic built environments in another key way: they have added a layer of awe-inspiring sound.

The sound track isn't just part of the background of playing; it's a major component of the gaming experience—particularly in the case of *Halo* and its famously spine-tingling score. Tracks on the *Halo 3* sound track have names like "Honorable Intentions," "This Is the Hour," and "Never Forget." Perhaps my favorite track is called simply "Ambient Wonder," a name that perfectly sums up the purpose of an epic environment: to create a space that completely absorbs and envelops the player in a sense of awe and wonder.

Halo's audio director, Martin O'Donnell, describes his goal in creating the score: "The music should give a feeling of importance, weight, and sense of the 'ancient' to the visuals of *Halo*." The score includes Gregorian chanting, a string orchestra, percussion, and Qawwali vocals, a Sufi devotional style of music intended to produce an ecstatic state in the listener.[27] These are timeless musical techniques for provoking our bodies' epic emotions—and video games increasingly make use of them. As one *Halo* player explains, "A great video game will make the hairs stand up on the back of your neck. Goose pimples will erupt. That tingly sensation overtakes your gut. It happens to me whenever I hear the *Halo* sound track."[28]

SO WHAT DO all of these extreme visual and audio environments add up to? Epic projects: collaborative efforts to tell stories and accomplish missions at extreme scales.

Epic environments inspire us to undertake epic projects, because they are a tangible demonstration of what is humanly possible when we all work together. Indeed, they *expand* our notion of what is humanly possible. And that's why exploring an epic environment like *Halo 3* inspires the kind of emotions that lead to large-scale cooperation, an epic achievement in and of itself.

Games journalist Margaret Robertson reflects, "*Halo* has always been a place where I feel good. I don't mean that in a James Brown sense. I mean it's a place where I feel virtuous. . . . [It] engenders a sense of honour and duty which actually make you feel like a better person. . . . What's the point of going to a better place if you aren't going to be a better person?"[29]

Epic Projects

While reaching the 10 billion kill milestone was a significant community achievement, *Halo* players have actually spent more time working on two other epic projects—both collaborative knowledge projects. The first epic project involves documenting the *Halo* world on wikis and discussion forums.

The other is a project to build up each other's collective ability to fight the Covenant and play a better game with each other. Both projects take place largely on discussion forums and wikis.

To give you an indication of the scale of the collective effort to document the *Halo* world and improve player ability within it, players have written more than 21 million discussion forum posts on the official Bungie *Halo* forums alone. Meanwhile, the largest *Halo* wiki has just under six thousand different articles, created and edited by 1.5 million registered users.

Halo players are also sharing knowledge to make each other better gamers. While the Halopedia wiki helps players construct the epic saga of the *Halo* series, the Halowiki (which describes itself as a "sister site" to Halopedia) focuses exclusively on multiplayer strategy and techniques. Its "values" statement sets the tone for epic knowledge sharing:

> This site serves one purpose: Halowiki.net shall help players at all skill levels improve and/or find even more enjoyment in their *Halo* 3 online experience. Share what you know. Let others share what they know with you. We must get even the most skilled players to share their knowledge. The end result shall be that we all raise our skills and fun together. Let's try to visit the limits of our abilities![30]

The scope of Halowiki is as staggering as its sister site. Under the tips section alone, there is an A to Z catalog of more than 150 different categories of tips, from "Bad habits to avoid in team games" and "Communication tips," to "How to use vehicles effectively" and "Last-resort tips when all else fails." Each individual category of tips contains hundreds of specific pieces of advice, contributed by different gamers.

The strategy section, on the other hand, contains more complex advice, sorted into roughly one hundred different categories, from "Close-range weapon mastery" and "Using ancient practices—advice from Sun Tzu's *The Art of War*," to one of my personal favorites: "Retraining your brain to not be afraid to die in the game."

In total, there are more than one thousand different sections on Halowiki that compile players' firsthand knowledge playing the game into a collective intelligence resource. Ultimately, for members of the *Halo* community, this resource serves a greater purpose than just creating better *Halo* players. Adding a bit of knowledge to the wiki validates that you know at least one thing that matters to millions of other people. It might be just a bit of *Halo* trivia — but there's nothing trivial about the positive feeling you get when you make a contribution that millions of other people can value and appreciate.

HALO **HAS CONSISTENTLY** pushed the limits of epic game design for a decade now — the first game in the series was released in 2001. But plenty of other online games are doing their part to invent new ways for gamers to become a part of something bigger. One of the most interesting recent experiments in epic game design is a project called Season Showdown, developed by EA Sports for its best-selling college football series *NCAA Football*. Season Showdown is the first significant effort in the sports video game genre — a highly successful genre, representing more than 15 percent of all game sales — to create the same kinds of epic emotional rewards more traditionally associated with save-the-world games like *Halo*.

"Every Game Counts" is the tagline for *NCAA Football 10*. Of course, this begs the question: counts toward what? The short answer is: every game played counts toward a national championship. It's not the real national championship, but not an entirely virtual one, either.

When you sign up to play *NCAA Football 10* online, the first thing you do is declare a team allegiance. You can pick any one of the 120 real-world college teams represented in the game, from Ohio State, Notre Dame, or Stanford to Florida State, Army, or USC. (I picked my alma mater, California.) For the rest of the online football season, every online point you score in the video game gets added to your team's score. The team scores are tallied weekly, in order to determine the winner in a series of school vs. school matchups.

These matchups perfectly mirror the real-world NCAA schedule. So, for example, the week that Oregon State faces Stanford in the real world, the two teams' fans will compete online in five head-to-head video game challenges. The team that carries three of the five challenges is crowned the week's online winner, regardless of who wins in the real world. That means plenty of online upsets, as fans of struggling teams rally to offset real losses with virtual victories.

At the end of the year, the best-performing online teams compete in their own conference championships. The ultimate payoff is an *NCAA Football 10* National Championship video game played out the same week as the real-world National Championship game. In the words of EA Sports, "The national champion will be composed of the most dedicated fans playing *NCAA 10*."

And that's what makes every *NCAA Football 10* game more meaningful than other sports video games. You're not just playing for yourself and for your own enjoyment. You're publicly playing to show support for your real favorite team, as part of a collective, fan-wide effort.

What's so innovative about *NCAA Football 10* is the fact that the game is using reality itself as the larger context for individual player actions. It's a fantasy league, but it's a fantasy league wrapped in reality. It doesn't have to invent a context from scratch to connect players to an epic story. Instead, it taps into existing college football narratives and traditions. It leverages existing communities, or fan bases, to provide meaningful social context. It feels epic because it's directly connecting fans to a much bigger organization they care deeply about, but can't ordinarily participate in directly.

As much fun as it is to cheer on our favorite teams, it's more meaningful to do something that pushes us to the edge of our own ability—and that counts, measurably. In *NCAA Football 10*, you're not just playing *as* your favorite college team, you're playing *in service of* your favorite college team. You're actively contributing to their reputation in a way that is quantified and amplified by EA Sports. As one blogger puts it, "Every game you play will help your school's cause."[31] It's all about being of service to a larger cause—one you already care about.

JOHAN HUIZINGA, the great twentieth-century Dutch philosopher of human play, once said, "All play *means* something."[32] Today, thanks to the increased scale of game worlds and advances in collective game design, gameplay often means something *more*. Game developers today are honing their ability to create awe-inspiring contexts for collective effort and heroic service. As a result, game communities are more committed than ever to setting extreme-scale goals and generating epic meaning.

When our everyday work feels trivial, or when we can't easily be of direct service to a larger cause, games can fulfill an important need for us. As we play games at an epic scale, we're increasing our ability to rise to the occasion, to inspire awe, and to take part in something bigger than ourselves.

Earlier in this chapter I quoted a *Halo* player who wondered, "Imagine what we could do with the full force of six billion humans!!"

Of course, there aren't enough Xboxes in the world to do it. Nor could everyone afford them, of course. But it does make for an interesting thought experiment: What *could* you do in a game like *Halo 3* if you had the full force of humanity playing together?

On one hand, this is an absurd idea to even consider. What would be the point of assembling 6 billion people to wage a fictional war?

But on the other: can you imagine what it would feel like to have 6 billion people fighting *on the same side* of a fictional war?

I think it's pretty clear that such an effort would have real meaning, even if it failed to generate any real-world value. If you were able to focus the attention of the entire planet on a single goal, even if just for one day, and even if it just involved dispatching aliens in a video game, it would be a truly awe-inspiring occasion. It would be the single biggest collective experience ever undertaken in the whole of human history. It would give the whole earth goose bumps.

That's the epic scale that gamers are capable of thinking on. That's the scale gamers are ready to work at.

Gamers can imagine 6 billion people coming together to fight a fictional enemy, for the sheer awe and wonder of it. They are ready to work together on extreme scales, toward epic goals, just for the spine-tingling joy of it. And the more we seek out that kind of happiness as a planet, the more likely we are to save it — not from fictional aliens, but from apathy and wasted potential.

Jean M. Twenge, a professor of psychology and the author of *Generation Me*, has persuasively argued that the youngest generations today — particularly anyone born after 1980 — are, in her words, "more miserable than ever before." Why? Because of our increased cultural emphasis on "self-esteem" and "self-fulfillment." But real fulfillment, as countless psychologists, philosophers, and spiritual leaders have shown, comes from fulfilling commitments to others. We want to be esteemed in the eyes of others, not for "who we are," but rather for what we've done that really matters.

The more we focus on ourselves and avoid a commitment to others, Twenge's research shows, the more we suffer from anxiety and depression. But that doesn't stop us from trying to make ourselves happy alone. We mistakenly think that by putting ourselves first, we'll finally get what we want. In fact, true happiness comes not from thinking *more* of ourselves, but rather from thinking *less* of ourselves — from seeing the truly small role we play in something much bigger, much more important than our individual needs.

Joining any collective effort and embracing feelings of awe can help us unlock our potential to lead a meaningful life and to leave a meaningful mark on the world.

Even if it's a virtual world we're leaving our mark on, we're still learning what it feels like to be of service to a larger cause. We're priming our brains and bodies to value and to seek out epic meaning as an emotional reward. And as recent research suggests, the more we enjoy these rewards in game worlds, the more likely we may be to seek them out in real life.

Three scientific studies published in 2009 by a consortium of researchers from eight universities in the United States, Japan, Singapore, and Malaysia studied the relationship between time spent playing games that require "helpful behavior" and the gamers' willingness to help others in everyday life. One study focused on children age thirteen and younger, another on teenagers, and the third on college students. The researchers worked with more than three thousand young gamers in total, and in all three studies they reached the same conclusion: young people who spend more time playing games in which they're required to help each other are significantly more likely to help friends, family, neighbors, and even strangers in their real lives.[33]

Although these studies weren't specifically looking at epic-scale games, the core findings seem likely to remain consistent, or even increase, at larger scales. As Brad Bushman, one coauthor of the studies and a professor of communications and psychology at the University of Michigan's Institute for Social Research, puts it, "These findings suggest there is an upward spiral of prosocial gaming and helpful behavior."[34] In other words, the more we help in games, the more we help in life. And so there's good reason to believe that the more we learn to enjoy serving epic causes in game worlds, the more we may find ourselves contributing to epic efforts in the real world.

THE PSYCHOLOGIST Abraham Maslow famously said, "It isn't normal to know what we want. It is a rare and difficult psychological achievement."[35] But today's best games give us a powerful tool for achieving exactly that rare kind of self-knowledge.

Games are showing us exactly what we want out of life: more satisfying work, better hope of success, stronger social connectivity, and the chance to be a part of something bigger than ourselves. With games that help us generate these four rewards every day, we have unlimited potential to raise our own quality of life. And when we play these games with friends, family, and neighbors, we can enrich the lives of people we care about.

So games are teaching us to see what really makes us happy—and how to become the best versions of ourselves. But can we apply that knowledge to the real world?

By supporting our four essential human cravings, and by providing a reliable source of flow and fiero, the gaming industry has gone a long way toward making us happier and more emotionally resilient—but only up to a point. We haven't learned how to enjoy our *real lives* more thoroughly. Instead, we've spent the last thirty-five years learning to enjoy our *game lives* more thoroughly.

Instead of fixing reality, we've simply created more and more attractive alternatives to the boredom, anxiety, alienation, and meaninglessness we run up against so often in everyday life. It's high time we start applying the lessons of games to the design of our everyday lives. We need to engineer *alternate* realities: new, more gameful ways of interacting with the real world and living our real lives.

Fortunately, the project of making alternate realities is already under way.

PART TWO

Reinventing Reality

All life is an experiment. The more experiments you make, the better.

—RALPH WALDO EMERSON

CHAPTER SEVEN

The Benefits of Alternate Realities

Whenever I walk through the front door of my apartment, I enter an alternate reality. It looks and works just like regular reality, with one major exception: when I want to clean the bathroom, I have to be *really* sneaky about it.

If my husband, Kiyash, thinks I'm going to scrub the tub on Saturday morning, he'll wake up early, tiptoe out of the bedroom and silently beat me to it. But I've lived in this alternate reality long enough to have developed a highly effective counterstrategy: I clean the bathroom at odd hours in the middle of the week, when he's least expecting it. The more random the hour, the more likely I am to complete the chore before he does. And if this strategy ever starts to fail? Well, let's just say that I am not above hiding the toilet brush.

Why exactly are we competing with each other to do the dirty work? We're playing a free online game called Chore Wars. And it just so happens that ridding our real-world kingdom of toilet stains is worth more experience points, or XP, than any other chore in the Land of the 41st-Floor Ninjas, which is what we've dubbed our apartment in the game. (We live on the forty-first floor, and my husband has a thing for *ninjutsu*.)

Chore Wars

Chore Wars is an alternate reality game (ARG), a game you play in your real life (and not a virtual environment) in order to enjoy it more. Chore Wars is essentially a simplified version of *World of Warcraft*, with one notable exception: all of the online quests correspond with real-world cleaning tasks, and instead of playing with strangers or faraway friends online, you play the game with your roommates, family, or officemates. Kevan Davis, a British experimental game developer who created Chore Wars in 2007, describes it as a "chore management system."[1] It's meant to help you track how much housework people are doing—and to inspire everyone to do more housework, more cheerfully, than they would otherwise.

To play Chore Wars, you first have to recruit a "party of adventurers" from your real-life household or office. That means getting your roommates, family members, or coworkers to sign up online, where together you'll name your kingdom and create avatars to represent everyone in the game.

Anyone who creates an avatar is eligible to undertake any of the custom "adventures" that you create in the game's database—in my household, these include emptying the dishwasher and brewing the first pot of coffee. And because it's a role-playing game, you're encouraged to write up the chores with a fantastical spin. In the Land of the 41st-Floor Ninjas, for example, brushing out our Shetland sheepdog is "Saving the dog-damsel in distress from clumps and shedding," and doing the laundry is "Conjuring clean clothes."

Whenever you complete one of these chores, you log in to the game to report your success. Every chore grants you a customized amount of experience points, virtual gold, treasure, avatar power-ups, or points that increase your virtual skills and abilities: plus ten dexterity points for dusting without knocking anything off the shelves, for example, or plus five stamina points for taking out all three kinds of recycling. And because you get to craft the adventures from scratch yourself, you can customize the in-game rewards to make the least popular chores more attractive—hence, the battle in my apartment to clean the bathroom first. It's worth a whopping one hundred XP.

The more chores you finish, the more experience points and virtual gold you earn, and the faster you level up your online avatar's powers. But Chore Wars isn't just about tracking your avatar development; it's also about earning real rewards. The game's instructions encourage households to invent creative ways to redeem the virtual gold in real life. You could exchange the gold for allowances if you're playing with your kids, or for rounds of drinks for roommates, or coffee runs for workmates, for example. My husband and I share a single car, so we use our gold pieces to bid on what music to play in the car whenever we're driving somewhere together.

But even more satisfying than all of my avatar powers, accumulated gold, and music privileges is the fact that after nine months of playing Chore Wars together, my husband's avatar has earned more overall experience points than I have. And avatar stats don't lie: for nearly a year now, Kiyash has definitely put in more effort cleaning the apartment than I have.

Clearly, this is a game that you win even if you lose. Kiyash has the satisfaction of being the best ninja on the forty-first floor, and I have the pleasure of doing fewer chores than my husband—at least until my competitive spirit kicks back in. Not to mention, it's more enjoyable to be partners in crime when it comes to housework, instead of nagging each other about chores. And, of course, as an added bonus, our place is cleaner than it ever has been before. Chore Wars has transformed something we both normally hate doing into something that feels creative and fun. The game has changed our reality of having to do housework, and for the better.

We're not alone. Chore Wars is one of the best reviewed and most beloved, if little known, secrets on the Internet.

A mom in Texas describes a typical Chore Wars experience: "We have three children, ages nine, eight, and seven. I sat down with the kids, showed them their characters and the adventures, and they literally jumped up and ran off to complete their chosen tasks. I've never seen my eight-year-old son make his bed! And I almost fainted when my husband cleaned out the toaster oven."

The experience apparently works as well for twentysomethings as it does for kids. As another player reports: "I live in a house in London with one other girl and six guys. A lot of the time I'm the only one tidying up, which was driving me slowly insane. I set up an account for us last night, and set some 'adventures,' and when I got up this morning *everyone in the house was cleaning*. I honestly could not believe what I was seeing. All we had to do is make it a competition! Now the guys are obsessed with beating each other!"[2]

How, exactly, does Chore Wars do it?

We typically think of chores as things we have to do. Either someone is nagging us to do them or we do them out of absolute necessity. That's why they're called chores: by definition, unpleasant tasks. The brilliant masterstroke of Chore Wars is that it convinces us that we *want* to do these tasks.

More important, however, is the introduction of *meaningful choice* into the housework equation. When you set up your party, your first task is to create a large pool of adventures to choose from. No player is assigned a particular adventure. Instead, everyone gets to pick their own. There are no *necessary* chores. You are volunteering for every adventure you take. And this sense of voluntary participation in housework is strengthened by the fact that you're encouraged to apply strategy as you choose your own housework adventures. Should you go for lots of chores that are fast and easy to complete, and try to rack up as many XP as possible that way? Or should you go for the harder, bigger chores, blocking other players from getting all that gold?

Of course, there are no good unnecessary obstacles without arbitrary restrictions. And for advanced Chore Wars players, that's where the real fun comes in. You can make it harder to earn XP and gold by adding new rules to any adventure. For example, you can set target time limits: double XP if you can put away your laundry in under five minutes. Or you can add a stealth requirement: you must empty the trash without anyone seeing you. Or you can simply tack on absurd restrictions: this chore must be done while singing, loudly, for example, or while walking backward.

It sounds ridiculous—why would making a chore harder make it more fun? But like any good game, the more interesting the restrictions, the more we enjoy playing. The Chore Wars management system makes it easy for players to dream up and try out new ways of doing the most ordinary things. Chores are, again by definition, routine—but they don't have to be. Doing them in a game format makes it possible to experience fiero doing something as mundane as cleaning up a mess, simply by making it more challenging, or by requiring us to be more creative about how we do it.

In real life, if you do your chores, there are visible results—a sparkling kitchen, or an organized garage. That's one kind of feedback, and it can certainly be satisfying. But Chore Wars smartly augments this small, everyday satisfaction with a more intense kind of feedback: avatar improvements. As online role-playing gamers everywhere know, leveling up is one of the most satisfying kinds of feedback ever designed. Watching your avatar profile get more powerful and skillful with each chore makes the work feel personally satisfying in a way that a cleaner room just doesn't. You are not just doing all this work for someone else. You are developing your own strengths as you play.

Best of all, you are getting better and better all the time. Even as the laundry gets dirty again or the dust starts to sneak back in, your avatar is still getting stronger, smarter, swifter. In this way, Chore Wars brilliantly reverses the most demoralizing aspects of regular housework. The results of a chore well done may start to fade almost immediately, but no one can take away the XP you have earned.

Individual success is always more rewarding when it happens in a multiplayer context, and this is part of Chore Wars' successful design as well. The game connects all of my individual activities to a larger social experience: I'm never just doing "my" chores; I'm playing with and competing against others. I can see how I measure up to others and compare avatar strengths to learn more about what makes me unique. Meanwhile, as I'm working, I'm thinking about the positive social feedback I'll get in the comments on my adventure, whether it's friendly taunts from a rival or OMGs of amazement for getting such a herculean task done.

Chore Wars isn't the kind of game you'd want to play forever; like all good games, their destiny is to become boring eventually, the better you get at them. But even if household interest in the game dies down after a few weeks or months, a major feat has been accomplished: players have had a rather memorable, positive experience of doing chores together. And that should change the way they think about and approach chores for some time.

So that's how Chore Wars achieves the seemingly impossible. It turns routine housework into a collective adventure, by adding unnecessary obstacles and implementing more motivating feedback systems. And it's the perfect example of our next reality fix:

FIX #7: WHOLEHEARTED PARTICIPATION

> Compared with games, reality is hard to get into. Games motivate us to participate more fully in whatever we're doing.

To participate wholeheartedly in something means to be *self-motivated* and *self-directed, intensely interested* and *genuinely enthusiastic*.

If we're forced to do something, or if we do it halfheartedly, we're not really participating.

If we don't care how it all turns out, we're not really participating.

If we're passively waiting it out, we're not really participating.

And the less we fully participate in our everyday lives, the fewer opportunities we have to be happy. It's that plain and simple. The emotional and social rewards we really crave require active, enthusiastic, self-motivated participation. And helping players participate more fully in the moment, instead of trying to escape it or just get through it, is *the* signature hallmark of alternate reality projects—the focus of this and the following three chapters of this book.

If "alternate reality" is an unfamiliar term for you, then you're not alone. Alternate reality development is still a highly experimental field. The term "alternate reality game" has been in use as a technical industry term since 2002, but there are still plenty of gamers and game designers who know little about it, let alone people outside of the gaming world.

As game developers are increasingly starting to push the limits of how much a game can affect our real lives, the concept of alternate reality is becoming more and more central to discussions about the future of games. It's helping to promote the idea that game technologies can be used to organize real-world activity. Most importantly, it's provoking innovative ideas about how to blend together what we love most about games and what we want most from our real lives.

On a recent Saturday morning, I found myself on Twitter, trading possible definitions for "alternate reality game" back and forth with about fifty other alternate reality gamers and developers. We were trying to work out a short definition that would really capture the spirit of ARG design, if not necessarily describe all the possible technological and formal components.

Collectively, we cobbled together a description of ARGs that seems to capture their spirit more effectively than any other definition I've seen: alternate realities are the *antiescapist* game.

ARGs are designed to make it easier to generate the four intrinsic rewards we crave — more satisfying work, better hope of success, stronger social connectivity, and more meaning — whenever we can't or don't want to be in a virtual environment. They're not meant to diminish the real rewards we get from playing traditional computer and video games. But they do make a strong argument that these rewards should be easier to get in real life.

In other words, ARGs are games you play to get more out of your real life, as opposed to games you play to escape it. ARG developers want us to participate as fully in our everyday lives as we do in our game lives.

Apart from this common mission, great alternate reality games can differ tremendously from one to another, in terms of style, scale, scope, and budget. Some ARGs, like Chore Wars, have relatively humble ambitions. They pick one very specific area of our personal lives and try to improve it. Others have quite audacious goals, involving entire communities or society at large: for example, to reinvent public education as we know it, to help players discover their true purpose in life, or even to improve our experience of death and dying.

Of course, not all ARGs are designed explicitly to improve our lives. Historically, in fact, most ARGs, like most computer and videogames, have been designed simply to be fun and emotionally satisfying. But my research shows that because ARGs are played in real-world contexts, instead of in virtual spaces, they almost always have at least the *side effect* of improving our real lives.[3] And so while others might distinguish between "serious" ARGs and "entertainment" ARGs, I prefer to look at *all* ARGs as having the potential to improve our quality of life. Indeed, a significantly higher percent of newer ARGs (created since 2007, compared with early ARGs created 2001–2006) are designed with explicit quality of life or world-changing goals. You'll read about these "positive impact" ARGs in the chapters ahead.

Some ARGs are invented and playtested on a shoestring budget, whether by artists, researchers, indie game developers, or nonprofit organizations. They're often developed for relatively small groups: a few hundred or a few thousand players. Others are backed by multimillion-dollar investments, receive funding from major foundations, or are sponsored by Fortune 500 companies. These bigger games can attract tens of thousands, hundreds of thousands, or even, in a few extremely successful cases, millions of players.[4]

Still, for the most part, alternate reality games today are small-scale probes of the future. They're a showcase for new possibilities. No single ARG is changing the world yet. But taken together, they're proving one at a time the myriad and important ways we could make our real lives better by playing more games.

So let's look at a few groundbreaking alternate reality projects. As we do, you'll notice that there are two key qualities that every good ARG shares.

First and foremost, like any good game, an ARG must always be *optional*. You can bet that if you *required* someone to play Chore Wars, it would lose a large part of its appeal and effectiveness. An alternate reality game has to remain a true "alternate" for it to work.

It's not enough, however, just to make something optional. Once the activity is under way, a good ARG, like any good game, also needs compelling goals, interesting obstacles, and well-designed feedback systems. These three elements encourage fuller participation by tapping into our natural desires to master challenges, to be creative, to push the limits of our abilities. And that's where **optimal experience design** comes in. Without a doubt, some alternate realities are more fun and engaging than others, just as some traditional games are better than others. The best ARGs are the ones that, like the best traditional computer and video games, help us create more satisfying work for ourselves, cultivate better hopes of success, strengthen our social bonds and activate our social networks, and give us the chance to contribute to something bigger than ourselves.

One ARG that achieves all of these goals is Quest to Learn—a bold new design for public schools that shows us how education can be transformed to engage students as wholeheartedly as their favorite video games.

Quest to Learn—And Why Our Schools Should Work More Like a Game

Today's "born-digital" kids—the first generation to grow up with the Internet, born 1990 and later—crave gameplay in a way that older generations don't.

Most of them have had easy access to sophisticated games and virtual worlds their entire lives, and so they take high-intensity engagement and active participation for granted. They know what extreme, positive activation feels like, and when they're not feeling it, they're bored and frustrated.[5] They have good reason to feel that way: it's a lot harder to function in low-motivation, low-feedback, and low-challenge environments when you've grown up playing sophisticated games. And that's why today's born-digital kids are suffering more in traditional classrooms than any previous generation. School today for the most part is just one long series of *necessary* obstacles that produce negative stress. The work is mandatory and standardized, and failure goes on your permanent record. As a result, there's a growing disconnect between virtual environments and the classroom.

Marc Prensky, author of *Teaching Digital Natives*, describes the current educational crisis:

> "Engage me or enrage me," today's students demand. And believe me, they're enraged. All the students we teach have something in

their lives that's really engaging—something that they do and that they are good at, something that has an engaging, creative component to it. . . . Video games are the epitome of this kind of total creative engagement. By comparison, school is so boring that kids, used to this other life, can't stand it. And unlike previous generations of students, who grew up without games, they know what real engagement feels like. They know exactly what they're missing.[6]

To try to close this gap, educators have spent the past decade bringing more and more games into our schools. Educational games are a huge and growing industry, and they're being developed to help teach pretty much any topic or skill you could imagine, from history to math to science to foreign languages. When these games work—when they marry good game design with strong educational content—they provide a welcome relief to students who otherwise feel underengaged in their daily school lives. But even then, these educational games are at best a temporary solution. The engagement gap is getting too wide for a handful of educational games to make a significant and lasting difference over the course of a student's thirteen-year public education.

What *would* make the difference? Increasingly, some education innovators, including Prensky, are calling for a more dramatic kind of game-based reform. Their ideal school doesn't *use* games to teach students. Their ideal school *is* a game, from start to finish: every course, every activity, every assignment, every moment of instruction and assessment would be designed by borrowing key mechanics and participation strategies from the most engaging multiplayer games. And it's not just an idea—the game-reform movement is well under way. And there's already one new public school entirely dedicated to offering an alternate reality to students who want to game their way through to graduation.

Quest to Learn is a public charter school in New York City for students in grades six through twelve. It's the first game-based school in the world—but its founders hope it will serve as a model for schools worldwide.

Quest opened its doors in the fall of 2009 after two years of curriculum design and strategic planning, directed by a joint team of educators and professional game developers, and made possible by funding from the MacArthur Foundation and the Bill and Melinda Gates Foundation. It's run by principal Aaron B. Schwartz, a graduate of Yale University and a ten-year veteran teacher and administrator in the New York City Department of Education. Meanwhile, the development of the school's curriculum and schedule has been led by Katie Salen, a ten-year veteran of the game industry and a leading researcher of how kids learn by playing games.

In many ways, the college-preparatory curriculum is like any other school's—the students learn math, science, geography, English, history, foreign languages, computers, and arts in different blocks throughout the day. But it's how they learn that's different: students are engaged in gameful activities from the moment they wake up in the morning to the moment they finish up their final homework assignment at night. The schedule of a sixth-grader named Rai can help us better understand a day in the life of a Quest student.

7:15 a.m. Rai is "questing" before she even gets to school. She's working on a secret mission, a math assignment that yesterday she discovered hidden in one of the books in the school library. She exchanges text messages with her friends Joe and Celia as soon as she gets up in order to make plans to meet at school early. Their goal: break the mathematical code before any of the other students discover it.

This isn't a mandatory assignment—it's a secret assignment, an opt-in learning quest. Not only do they not have to complete it, they actually have to *earn the right* to complete it, by discovering its secret location.

Having a secret mission means you're not learning and practicing fractions because you have to do it. You're working toward a self-chosen goal, and an exciting one at that: decoding a secret message before anyone else. Obviously not all schoolwork can be special, secret missions. But when every book could contain a secret code, every room a clue, every handout a puzzle, who wouldn't show up to school more likely to fully participate, in the hopes of being the first to find the secret challenges?

9:00 a.m. In English class, Rai isn't trying to earn a good grade today. Instead, she's trying to level up. She's working her way through a storytelling unit, and she already has five points. That makes her just seven points shy of a "master" storyteller status. She's hoping to add another point to her total today by completing a creative writing mission. She might not be the first student in her class to become a storytelling master, but she doesn't have to worry about missing her opportunity. As long as she's willing to tackle more quests, she can work her way up to the top level and earn her equivalent of an A grade.

Leveling up is a much more egalitarian model of success than a traditional letter grading system based on the bell curve. Everyone can level up, as long as they keep working hard. Leveling up can replace or complement traditional letter grades that students have just one shot at earning. And if you fail a quest, there's no permanent damage done to your report card. You just have to try more quests to earn enough points to get the score you want. This system of "grading" replaces negative stress with positive stress, helping students focus more on learning and less on performing.

11:45 a.m. Rai logs on to a school computer to update her profile in the "expertise exchange," where all the students advertise their learning superpowers. She's going to declare herself a master at mapmaking. She didn't even realize mapmaking could count as an area of expertise. She does it for fun, outside of school, making maps of her favorite 3D virtual worlds to help other players navigate them better. Her geography teacher, Mr. Smiley, saw one of her maps and told her that eighth-graders were just about to start a group quest to locate "hidden histories" of Africa: they would look for clues about the past in everyday objects like trade beads, tapestries, and pots. They would need a good digital mapmaker to help them plot the stories about the objects according to where they were found, and to design a map that would be fun for other students to explore.

The expertise exchange works just like video game social network profiles that advertise what games you're good at and like to play, as well as the online matchmaking systems that help players find new teammates. These systems are designed to encourage and facilitate collaboration. By identifying your strengths and interests publicly, you increase the chances that you'll be called on to do work that you're good at. In the classroom, this means students are more likely to find ways to contribute successfully to team projects. And the chance to do something you're good at as part of a larger project helps students build real esteem among their peers — not empty self-esteem based on nothing other than wanting to feel good about yourself, but actual respect and high regard based on contributions you've made.

2:15 p.m. On Fridays, the school always has a guest speaker, or "secret ally." Today, the secret ally is a musician named Jason, who uses computer programs to make music. After giving a live demonstration with his laptop, he announces that he'll be back in a few weeks to help the students as a coach on their upcoming "boss level." For the boss level, students will form teams and compose their own music. Every team will have a different part to play — and rumor has it that several mathematical specialists will be needed to work on the computer code. Rai really wants to qualify for one of those spots, so she plans to spend extra time over the next two weeks working harder on her math assignments.

As the Quest website explains, boss levels are "two-week 'intensive' [units] where students apply knowledge and skills to date to propose solutions to complex problems." "Boss level" is a term taken directly from video games. In a boss level, you face a boss monster (or some equivalent thereof) — a monster so intimidating it requires you to draw on everything you've learned and mastered in the game so far. It's the equivalent of a midterm or final exam. Boss levels are notoriously hard but immensely satisfying to beat. Quest schedules boss levels at various points in the school year, in order to fire students up about putting their lessons into action. Students get to tackle an epic challenge — and there's no shame in failing. It's a boss level, and so, just like any good game, it's meant to whet your appetite to try harder and practice more.

Like collaborative quests, the boss levels are tackled in teams, and each student must qualify to play a particular role — "mathematical specialist," for example. Just as in a big *World of Warcraft* raid, each participant is expected to play to his or her strengths. This is one of Quest's key strategies for giving students better hopes of success. Beyond the basic core curriculum, students spend most of their time getting better at subjects and activities — ones they have a natural talent for or already know how to do well. This strategy means every student is set up to truly excel at something, and to focus attention on the areas in which he or she is most likely to one day become extraordinary.

6:00 p.m. Rai is at home, interacting with a virtual character named Betty. Rai's goal is to teach Betty how to divide mixed numbers. Betty is what Quest calls a "teachable agent": "an assessment tool where kids teach a digital character how to solve a particular problem." In other words, Betty is a software program designed to know *less* than Rai. And it's Rai's job to "teach" the program, by demonstrating solutions and working patiently with Betty until she gets it.

At Quest, these teachable agents replace quizzes, easing the anxiety associated with having to perform under pressure. With a teachable agent, you're not being tested to see if you've really learned something. Instead, you're mentoring someone because you really have learned something, and this is your chance to show it. There's a powerful element of naches—vicarious pride—involved here: the more a student learns, the more he or she can pass it on. This is a core dynamic of how learning works in good video games, and at Quest it's perfectly translated into a scalable assessment system.

Secret missions, boss levels, expertise exchanges, special agents, points, and levels instead of letter grades—there's no doubt that Quest to Learn is a different kind of learning environment, about as radically different a mission as any charter school has set out in recent memory. It's an unprecedented infusion of gamefulness into the public school system. And the result is a learning environment where students get to share secret knowledge, turn their intellectual strengths into superpowers, tackle epic challenges, and fail without fear.

Quest to Learn started with a sixth-grade class in the fall of 2009, and it plans to add a new sixth-grade class each year as the previous year graduates upward. The first senior class will graduate from Quest to Learn in 2016, and potentially from college by 2020. I'm willing to bet that that graduating class will be full of creative problem solvers, strong collaborators, and innovative thinkers ready to wholeheartedly tackle formidable challenges in the real world.

SuperBetter—Or How to Turn Recovery into a Multiplayer Experience

Either I'm going to kill myself or I'm going to turn this into a game. After the four most miserable weeks of my life, those seemed like the only two options I had left.

It was the summer of 2009, and I was about halfway through writing this book when I got a concussion. It was a stupid, fluke accident. I had been standing up, and I slammed my head straight into a cabinet door I didn't realize was still open. I was dizzy, saw stars, and felt sick to my stomach. When my husband asked me who the president was, I drew a blank.

Some concussions get better in a few hours, or a few days. Others turn into a much longer postconcussion syndrome. That's what happened to me. I got a headache and a case of vertigo that didn't go away. Any time I turned my head, it felt like I was doing somersaults. And I was in a constant mental fog. I kept forgetting things—people's names, or where I'd put things. If I tried to read or write, after a few minutes my vision blurred out completely. I couldn't think clearly enough to keep up my end of interesting conversations. Even just being around other people, or out in public spaces, seemed to make it worse. At the time, I scribbled these notes: "Everything is hard. The iron fist pushes against my thoughts. My whole brain feels vacuum pressurized. If I can't think, who am I?"

After five days of these symptoms and after a round of neurological tests that all proved normal, my doctor told me I would be fine—but it would probably take an entire month before I really felt like myself again. In the meantime, no reading, no writing, no working, and no running, unless I was completely symptom-free. I had to avoid anything that made my head hurt or made the fog worse. (Sadly, I quickly discovered that computer and video games were out of the question; it was way too much mental stimulation.)

This was difficult news to hear. A month seemed like an impossibly long time not to work and to feel this bad. But at least it gave me a target to shoot for. I set the date on my calendar: August 15, I would be better. I believed it. I *had* to believe it.

That month came and went, and I'd barely improved at all.

That's when I found out that if you don't recover in a month, the next likely window of recovery is three months.

And if you miss *that* target, the next target is a year.

Two more months living with a vacuum-pressurized brain? Possibly an *entire year*? I felt more hopeless than I could have ever imagined. Rationally, I knew things could be worse—I wasn't dying, after all. But I felt like a shadow of my real self, and I wanted so desperately to resume my normal life.

My doctor had told me that it was normal to feel anxious or depressed after a concussion. But she also said that anxiety and depression exacerbate concussion symptoms and make it much harder for the brain to heal itself. The more depressed or anxious you get, the more concussed you feel and the longer recovery takes. Of course, the worse the symptoms are and the longer they last, the more likely you are to be anxious or depressed. In other words, it's a vicious cycle. And the only way to get better faster is to break the cycle.

I knew I was trapped in that cycle. The only thing I could think of that could possibly make me optimistic enough to break it was a game.

It was a strange idea, but I literally had nothing else to do (except watch television and go on very slow walks). I'd never made a health care game before. But it seemed like the perfect opportunity to try out my alternate reality theories in a new context. I might not be able to read or write very much, but hopefully I could still be creative.

I knew right away it needed to be a multiplayer game. I'd been having a lot of trouble explaining to my closest friends and family how truly anxious I was and how depressed I felt, how hard the recovery process was. I also felt awkward, and embarrassed, asking for help. I needed a way to help myself tell my closest friends and family, "I am having the hardest time of my life, and I really need you to help me." But I also didn't want to be a burden. I wanted to *invite* people to help me.

As with any alternate reality project, I needed to research the reality of the situation before I could reinvent it. So, for a few days, I spent the limited amount of time I was able to focus—about an hour a day at that point—learning about postconcussion syndrome online. From various medical journals and reports, I pieced together what experts agree are the three most important strategies for getting better and coping more effectively—not only from concussions, but any injury or chronic illness.

First: stay optimistic, set goals, and focus on any positive progress you make. Second: get support from friends and family. And third: learn to read your symptoms like a temperature gauge. How you feel tells you when to do more, do less, or take breaks, so you can gradually work your way up to more demanding activity.[7]

Of course, it immediately occurred to me that these three strategies sound exactly like what you do when you're playing a good multiplayer game. You have clear goals; you track your progress; you tackle increasingly difficult challenges, but only when you're ready for them; and you connect with people you like. The only thing missing from these recovery strategies, really, was the meaning—the exciting story, the heroic purpose, the sense of being part of something bigger.

So that's where SuperBetter comes in.

SuperBetter is a superhero-themed game that turns getting better into multiplayer adventure. It's designed to help anyone recovering from an injury or coping with a chronic condition get better sooner—with more fun, and with less pain and misery, along the way.

The game starts with five missions. You're encouraged to do at least one mission a day, so that you've successfully completed them all in less than a week. Of course, you can move through them even faster if you feel up to it. Here are excerpts from the instructions for each mission, along with an explanation of how I designed it and how I played it.

> Mission #1: Create your SuperBetter secret identity. You're the hero of this adventure. And you can be anyone you want, from any story you love. So pick your favorite story—anything from James

Bond to *Gossip Girl*, *Twilight* to *Harry Potter*, *Batman* to *Buffy the Vampire Slayer*. You're about to borrow their superpowers and play the leading role yourself.

I chose *Buffy the Vampire Slayer* as my story line. That made me Jane the Concussion Slayer, and that made my symptoms the vampires, demons, and other forces of darkness I was destined by fate to battle against. The point of this mission is to start seeing yourself as powerful, not powerless. And it underscores the fact that you *are* heroic for choosing to persevere in the face of your injury or illness.

> Mission #2: Recruit your allies. Every superhero has an inner circle of friends who help save the day. Pick the people you want to count on most, and invite them to play this game with you. Ask each one to play a specific part: Batman needs a Robin and an Alfred, while James Bond needs an M, a Q, and a Moneypenny. If you're Bella, you'll want at least an Edward, a Jacob, and an Alice. Give each ally a specific mission, related to his or her character. Use your imagination—and feel free to ask for anything you need! When you're saving the world, you can't be shy about asking for help. Be sure to ask at least one ally to give you daily or weekly achievements—these are surprise accomplishments they bestow upon you based on your latest superheroic activities.

As Jane the Concussion Slayer, I recruited my twin sister as my "Watcher" (Buffy's mentor in the TV series). Her mission was to call me every single day and ask for a report on my concussion-slaying activities. She should also give me advice and suggest challenges for me to try. Before playing SuperBetter, I hadn't known how to explain to her that I really needed daily contact, and not just to hear from her on the weekends.

I recruited my husband as my "Willow" (Buffy's smarty-pants best friend who's also a computer geek). His mission was to do all of the score- and record-keeping for me, read me interesting articles, and in general help me with anything I wanted to do on the computer without getting a headache. Finally, I recruited my friends Natalie and Rommel, and their miniature dachshund, Maurice, as my "Xander" (he's the comic-relief character). Their mission was to come over once a week and just generally cheer me up.

Why recruit allies? Social psychologists have long observed that one of the hardest things about a chronic injury or illness is asking our friends and family for support. But reaching out and really asking for what we need makes a huge difference. It prevents social isolation, and it gives people who want to help, but don't know how, something specific and actionable to do.

And why have achievements? Every fiero moment helps increase optimism and a sense of mastery, which has been proven to speed recovery from everything from knee injuries to cancer. But achievements feel more meaningful when someone else gives them to you—that's why it's important to have a friend or family member bestow them upon you. Kiyash gave me my achievements based on the titles of episodes of *Buffy the Vampire Slayer*. (For example, I unlocked the "Out of Mind, Out of Sight" achievement for ignoring my e-mail for an entire day, and "The Harvest" achievement for eating vegetables for dinner instead of cookies and ice cream, which was one of my favorite postconcussión ways to drown my sorrows. At the time, both of those felt like epic struggles.)

> Mission #3: Find the bad guys. To win this battle, you need to know what you're up against. Pay attention all day to anything that makes you feel worse, and put it on your bad-guys list. Some days, you'll be able to battle the bad guys longer—some days not so long. But every time you do battle, you'll want to make a great escape. That means getting away from the bad guy before he knocks you flat. You can always add more bad guys to your list as you discover them—and if you vanquish one forever, you can take it off and claim the permanent victory.

My list of bad guys at the start of the game focused on activities I kept trying to sneak in even though I knew they made me feel worse: reading and responding to e-mail, running or doing any kind of vigorous exercise, playing *Peggle*, drinking coffee.

The better you can identify triggers of your symptoms, the more pain and suffering you'll avoid. And making a great escape turns a potential moment of failure—*This is harder than it should be*, or *I can't do what I want to do*—into a moment of triumph: *I succeeded in recognizing a trigger and vanquished it before it did too much damage.* One of the highlights in my recovery was when I enlisted the entire crew at the Peet's Coffee down the block to help me modulate the amount of caffeine in my morning iced coffee, which I was really reluctant to give up. It was their idea to start me off with 90 percent decaf with just a splash of caffeine so that I could work my way up to half and half, and eventually full caffeine when my brain was finally ready to be stimulated again.

> Mission #4: Identify your power-ups. Good thing you've got super-powers. Maybe they're not your typical superpowers—but you definitely have fun things you can do for yourself at a moment's notice to feel better. Make a list, and be ready to call on them whenever the bad guys are getting the better of you. In fact, try to collect as many power-ups as you can every day!

For my concussion recovery, I focused on things I could do with my senses that weren't affected by my head injury. Touch was fine, so I could sit and cuddle with my Shetland sheepdog. Hearing was fine, so I could sit by the window and listen to a podcast. And the biggest superpower I discovered had to do with my sense of smell: I really started to enjoy smelling different perfumes. I would go to a perfume counter, spray samples of a dozen perfumes on cards, then take them home and smell them throughout the rest of the evening, to see how they changed and to learn the different notes. It was one of the most engaging activities I could do without hurting my brain at all. And eventually, once my vertigo was improved, I was able to add to my power-up list long walks up San Francisco hills with my husband.

The power-ups are meant to help you feel capable of having a good day, no matter what. Having specific positive actions to take increases the odds of doing something that will break the cycle of feeling negative stress or depression.

> Mission #5: Create your superhero to-do list. Not every mission is possible, but it doesn't hurt to dream big. Make a list of goals for yourself, ranging from things you're 100 percent positive you can do right now to things you might not have been able to do even in your wildest dreams before you got sick or hurt. Everything on your list should be something that would make you feel awesome and show off your strengths. Every day, try to make progress toward crossing one of these superhero to-dos off your list. Be sure to get your allies' help and advice.

This final idea was inspired by a question I'd found on the website of a New Zealand occupational therapist. "If I can't take your pain away, what else would you like to improve in your life?"[8] It's one of the abiding features of a good game: the outcome is uncertain. You play in order to discover how well you can do—not because you're guaranteed to win. SuperBetter has to acknowledge the possibility of failure to achieve complete recovery. But it can also make it less scary to fail—because there is an abundance of other goals to pursue and other rewarding activities to undertake along the way. That's why it seemed essential to make part of the game a project to discover as many positive activities that it was still possible to do. It increased my real hopes of enjoying life more, no matter what else happened with the recovery or treatment.

One of my easiest superhero to-dos was baking cookies for people who live in my neighborhood. I liked it so much, I did it three times. A more challenging to-do was finding an opportunity to wear my favorite pair of purple leather stiletto boots, which meant getting up the energy to go out and see people. (I crossed this one off my list by going to see a movie with a big group of friends. I was a bit overdressed, but I felt great anyway.) The biggest superhero to-do on my list was, of course, to finish this book.

Once you have completed the five big missions, your challenge is to stay in constant contact with your allies, collect power-ups by battling the bad guys and making great escapes, and tackle items on your superhero to-do list. You might want to "lock in" your gameplay by keeping a game journal, posting daily videos on YouTube, or using Twitter to announce your achievements.

Near the end of every day, hold a secret meeting with one of your allies. Add up your great escapes, your power-ups, and your superhero points.

Talk to your other allies as often as possible, and tell them what you've been doing to get superbetter. Ask them for ideas about new things to add to your to-do list.

Be sure you have at least one ally who is giving you daily achievements. Share these achievements with your friends online, using Twitter or Facebook status updates, to keep them posted on your progress.

So that's how you play SuperBetter. But does it actually improve the reality of getting better?

The first few days I was playing, I was in a better mood than I had been at any time since I hit my head. I felt like I was finally *doing* something to get better, not just lying around and waiting for my brain to hurry up and heal itself.

My symptoms didn't improve instantly—but I was so much more motivated to get something positive out of my day, no matter what. Every day, no matter how bad I felt otherwise, I would score at least one great escape, grab at least one power-up, rack up some points, and unlock an achievement. Doing these things didn't require being cured; it just required making an effort to participate more fully in my own recovery process.

There's not a whole lot you can prove with a scientific sample of one. I can say only that, for me, the fog of misery lifted first, and then, soon after, the fog of symptoms started to lift as well. Within two weeks of playing Jane the Concussion Slayer, my symptoms were improved by roughly 80 percent, according to the log Kiyash helped me keep of my pain and concentration problems on a ten-point scale, and I was up to working as many as four hours a day. Within a month, I felt almost completely recovered.

I can't say for sure if I got better any faster than I would have without playing the game—although I suspect it helped a great deal. What I can say for sure is that I suffered a great deal less during the recovery as a direct result of the game. I was miserable one day, and the next day I wasn't; and I was never that miserable again as long as I was playing the game. When my allies joined the game, I finally felt like they really understood what I was going through, and I never felt quite so lost in the fog again.

After declaring my victory over the concussion in a Twitter post, I received dozens of requests to post all the rules and missions, so that other people could game their own injuries and illnesses—for everything from chronic back pain and social anxiety to lung disorders, migraines, the side effects of quitting smoking, newly diagnosed diabetes, chemotherapy, and even mononucleosis.

I published the rule set on my blog, and I gave it the more general name SuperBetter (after all, most people probably don't dream of being like Buffy the Vampire Slayer).[9] I suggested that people use the hashtag "#SuperBetter" for their own videos, blog posts, and Twitter updates, in case they wanted to find each other online. (A *hashtag* is a way to easily add context to your online content, and to find other people talking about the same topic.) And that was it. I didn't build a Web application, or develop an automated scoring system, or even set up a social network for playing the game. A game doesn't have to be a computer program. It can simply be like chess or hide-and-seek: a set of rules that one player can pass on to another.

An alternate reality game can be as simple as a good idea, a fresh way of looking at a problem. SuperBetter, of course, isn't meant to replace conventional medical advice or treatment. It's meant to augment good advice, and to help patients take a more active role in their own recovery.

When you're sick or in pain, getting better is all you want. But the longer it takes, the harder it gets. And when the tough reality we have to face is that getting better won't be easy, a good game can better prepare us to deal with that reality. In an alternate reality linked to our favorite superhero mythology, we're more likely to stay optimistic, because we'll set more reasonable goals and keep better track of our progress. We'll feel successful even when we're struggling, because our friends and family will define fiero moments for us every day. We'll build a stronger social support system, because it's easier to ask someone to play a game than it is to ask for help. And we'll hopefully find real meaning and develop real character in our epic efforts to overcome what may be the toughest challenge we've ever had to face. And *that's* how we get superbetter, thanks to a good game.

THE THREE GAMES discussed in this chapter represent three of the main approaches to developing an alternate reality and solving a quality-of-life problem.

Chore Wars is an example of a **life-management** ARG—a software program or service that helps you manage your real life like a game.

Quest to Learn is an example of an **organizational** ARG. It uses game design as a guiding philosophy for creating new institutions and inventing new organizational practices.

And SuperBetter is a **concept** ARG. It uses social media and networking tools to virally spread new game ideas, missions, and rule sets, which players can repurpose and adapt for their own lives as they see fit.

These three methods aren't the only ways to create an alternate reality. In later chapters in this book, you'll also read about **live event ARGs,** which gather players at physical locations for a game that takes only an hour or a day to play, and **narrative ARGs,** which use multimedia storytelling—video, text, photographs, audio, and even graphic novels—to weave real-world game missions into a compelling fiction that plays out over weeks, months, or even years.

Of course, by the time you read this book, dozens—probably hundreds—of new alternate reality games will no doubt be widely playable. This movement is just getting started. When we imagine how the ARG movement might unfold, we can—as always—look for guidance from the past.

In the early 1970s, just before the computer and video game revolution, another game revolution took place, with significantly less fanfare but a rather important and lasting legacy. It was called the New Games movement, and its goal was to reinvent sports to be more cooperative, more social, and more inclusive.

The New Games philosophy was simple, composed of two parts. First, no one should ever have to warm the bench because they're not good enough to play. And second, competitive gameplay shouldn't be about winning. It should be about playing harder and longer than the other team, in order to have more fun.

The founders of the movement, a group of San Francisco–based counterculturists, invented dozens of new sports, all sillier and more spectacular than traditional athletic activities. The most well known were the "earth ball" games (played with a ball six feet in diameter, so that it takes multiple people to move the ball together) and parachute games (in which twenty to fifty people stand around the rim of a piece of parachute material and flap and billow it together, working to create various shapes and ripples). They held large New Games festivals in the Bay Area and eventually trained tens of thousands of schools and parks and recreation departments across the country, so that they could include New Games in their physical education and public recreation programs.

Many of today's leading game developers grew up playing New Games at school and local parks—and it's not hard to see the influence of New Games on multiplayer and massively multiplayer game designers today. From the cooperative missions in MMOs to the 256-player combat environments on consoles, video gameplay today often looks a lot like a New Game, set in a virtual world. In fact, New Games theory has come up at every single Game Developers Conference I've attended over the last decade—which is how I know that many game designers have managed to acquire for themselves a copy of the long out-of-print and little-known *New Games Book*, published in 1976.

The New Games Book includes instructions for how to play the new sports and, more importantly, essays explaining the philosophy of the movement. Many of my friends in the industry have acknowledged they've flipped through its pages for game-design inspiration.

I've nearly worn the print off the page of my favorite essay in the book. It's called "Creating the Play Community," by Bernie DeKoven, then the codirector of the New Games Foundation and today a leading play theorist. In the essay, DeKoven calls for a community of players to volunteer to be of service to the movement. He asks: Who will be willing to try these new games and help assess whether they are, in fact, better than the old games? If they are better, the community should teach others how to play. If they're not better, the players should suggest ways to improve them, or start inventing their own new games to test. He explains:

> Because the games are new, we get a sense that we're experimenting. No one guarantees anything. If a game doesn't work, we try to fix it, to see if we can make it work. After all, it's a new game. It's not official yet. In fact, we're the officials, all of us, every one of us who has come to play. We make the judgments. We each take the responsibility for discovering what we can enjoy together.[10]

This is the kind of community that is currently coming together around alternate reality games. As we develop alternate realities, we need to be both open-minded and critical about what actually raises our quality of life, what helps us participate more fully in our real lives, and what simply serves as yet another distraction. There will be many, many different alternate realities proposed in the coming years, and it's not up to just the game developers to shape this movement. The players, more than anyone else, will get to decide if a new alternate reality is indeed a good game.

The "how" of alternate reality game design boils down to the game-design principles that best generate the four rewards we crave most. Traditional computer and video game developers are leading the way, constantly innovating new ways to reap these rewards; ARG developers are already borrowing and refining these design strategies and development tools as their go-to solutions for how to make the world work more like a game.

But as we playtest different possibilities to decide what makes a good alternate reality, three additional sets of criteria are certain to emerge.

First: *When* and *where* do we need an alternate reality? Which situations and spaces call for it—and when are we better off leaving reality alone?

Second: *Who* should we include in our alternate reality games? Besides our close friends and family, who else would we benefit from inviting to play with us?

And third: *What* activities should we be adopting as the core mechanics of our alternate reality games? Game design is a structure—goals, restrictions, feedback—but within that structure, we can ask players to do almost anything. What habits should we be encouraging? What actions should we be multiplying?

These three different sets of criteria are the subjects of the next three chapters, which in turn cover three key kinds of alternate reality projects: alternate realities designed to make difficult activities more rewarding, alternate realities designed to build up new real-world communities, and alternate realities designed to help us adopt the daily habits of the world's happiest people in our real, everyday lives.

CHAPTER EIGHT

Leveling Up in Life

HOW ALTERNATE REALITIES CAN MAKE
DIFFICULT ACTIVITIES MORE REWARDING

f I have one regret in life," I complained to the crowd at the Austin Convention Center, "it's that my undead priest is smarter than I am." Technically speaking, it's true: if you were to add up every A I've gotten in my real life, from junior high through graduate school, the total still wouldn't come close to my *World of Warcraft* character's intellect stat. Never mind the fact that there's no score at all for getting smarter once you're out of school for good.

I was giving a keynote address at the annual design and technology conference SXSW Interactive when I made this lament. The topic was, naturally, the failures of the real world to be as engaging as a good game, and what we could do to fix it. As I told the crowd, "I'd feel a lot better if I got plus-one intellect for every smart thing I said during this talk. Or at least a few plus-one public speaking points." Giving talks is exhausting, even when I enjoy it, I explained. It would be energizing to see some +1s pop up right on top of my PowerPoint slides as I worked my way through the deck.

A few days later, back home in California, I received an e-mail from an unfamiliar sender: ratings@plusoneme.com. The subject was "Clay has acknowledged your strengths." Clay who? I wondered. I didn't know anyone named Clay. I opened the e-mail anyway.

A friend of yours, Clay Johnson, +1d you to acknowledge some of your strengths. Specifically they're acknowledging these attributes:

+1 Intelligence
+1 Public Speaking
+1 Inspiration

Enjoy your day. And congratulations!

A second e-mail arrived a few minutes later, from Clay Johnson himself.

Your +1 in public speaking as you requested at SXSW! It should have arrived in your inbox a little while ago. When you said that during your speech, I thought, "Why shouldn't she be able to get a +1 in public speaking?!" and built plusoneme.com. Great talk. Check out what you inspired.

I followed the link, and sure enough, there was a perfect little Web application dedicated to giving and tracking stats in an array of thirty-seven different personal strengths: creativity, generosity, speed, fashion, listening, and backbone, for example.

It was definitely a broader and more diverse set of stats than I'd even seen in a role-playing game. For every plus-one you send, you can also attach a reason: "+1 backbone for sticking up for our idea in the meeting," for example, or "+1 endurance for getting through the long drive home tonight." And you can send a plus-one to anyone via e-mail, regardless of whether or not they've signed up to play. If they join the site and create a profile, their plus-ones "stack," or add up over time. (So far, I'm up to +25 innovation, because I asked my colleagues to plusoneme when I do something innovative at work.)

You can add a plus-one feed to your blog or social network page so that your friends and family can see exactly how fast you're leveling up, in what strengths.

All in all, Plusoneme is pretty much exactly what you'd wish for if you wanted to level up in real life—that is, if you wanted to have an objective measure of how much better you're getting at the things you're working hard at.

Since he gave me my first plus-one, I've gotten to know Clay Johnson better. It turns out that he's the director emeritus of Sunlight Labs, a community of open-source developers dedicated to making government more transparent and participatory. We've had some very interesting conversations about how to use game feedback systems to increase democratic participation. Frankly, I wouldn't be surprised to see a Plusoneme.gov someday, to help constituents give better feedback to their elected officials.

Plusoneme isn't a game—there aren't any built-in goals, and there are no restrictions on how you give or earn a plus-one. It's more like a *gesture* toward a game, a kind of musing out loud: How would it feel to get constant, real-time positive feedback in our real lives, whenever we're tackling obstacles and working hard? Would we be more motivated? Would we feel more rewarded? Would we challenge ourselves more?

A growing number of alternate reality projects suggest that, for all these questions, the answer is a resounding yes. Systems that help us *level up in real life*, by providing us with voluntary obstacles related to our real-world activity and by giving us better feedback really can help us make a better effort. And that gives us our next fix:

↻ FIX #8: MEANINGFUL REWARDS
WHEN WE NEED THEM MOST

Compared with games, reality is pointless and unrewarding.
Games help us feel more rewarded for making our best effort.

I hate flying, and I spend a *lot* of time hating it—on the order of over 150 hours a year.

I'm a nervous flier. I've gotten better over the years, but I still can't really work on planes, eat on planes, or sleep on planes. I certainly can't *enjoy* myself on planes. Half the time, I literally make myself sick with anxiety. Even after a good flight, I'm so exhausted from the stress and the jet lag that it takes hours or even a whole day or more to recover.

More than 25 million Americans have a fear of flying, while 52 percent of frequent fliers say that the number one word to describe air travel is "frustrating."[1] And this has significant consequences for our health and well-being.

Being out of control is a fundamentally stressful feeling. Researchers have shown that it takes a huge hit on both our happiness and our physical health. And it's not just in the moment that we're negatively affected. When we go through an experience that makes us feel endangered or powerless, our immune system suffers and we experience higher levels of anxiety, depression, and pessimism in the hours and days that follow.[2]

Games, of course, help put people back in control. Real gameplay is always by definition voluntary; it is always an exercise of our own freedom. Meanwhile, progressing toward goals and getting better at a game instills a sense of power and mastery.

The fact that commercial flying puts so many people on edge, so reliably, makes airports and airplanes the perfect target for a game-design intervention. If we could look forward to flights instead of dreading them, and if we could feel powerful at the start of our trip instead of helpless, the quality of life of frequent fliers worldwide would skyrocket. And the most fearful fliers would be able to go on more of the trips they want to take but currently avoid.

But what would make flying more authentically rewarding? Forget frequent-flier miles and other travel reward programs. If you're already frustrated or fearful about flying, earning more flights isn't going to make you any happier.

What we need are *intrinsic* reward programs—and two new games for fliers show exactly how it could be done: Jetset, the world's first video game for airports, and Day in the Cloud, an in-flight scavenger hunt designed to be played plane versus plane, at ten thousand feet and higher.

Jetset and Day in the Cloud

Jetset, an iPhone game created by Atlanta-based developers Persuasive Games, is a cartoon simulation of an airport security line. Load the game and, on your iPhone screen, you get to watch virtual passengers march through a cartoon metal detector while virtual luggage rolls through the X-ray machine. Your role in the game is to play the part of the security agent: tap the screen to confiscate banned items and to pat down suspicious passengers. Go too slow, and passengers miss their flights; go too quickly, and you might miss a banned item or let the wrong passenger slip by. The longer you play, the longer the line gets, the faster the security belt runs, and the harder it is to keep up with new security restrictions, like "no pressurized cheese," "no pet snakes," "no pudding cups," and "no robots."

The game's lead designer, Ian Bogost, is a frequent business traveler who came up with the idea for the game after suffering endless frustration in the security line himself. The game has a decidedly satirical bent, and player reviews often mention laughing out loud as they play.[3] That's one of the main goals of the game, Bogost told me: to make players laugh during a stressful situation. "Hopefully, it helps frequent fliers laugh at the absurdity of the airport security process instead of being overwhelmed or infuriated by it."

Technically, you can play Jetset anywhere you take your mobile phone. But the only way to officially level up and unlock souvenir prizes to send to friends and family is by playing the game at real-world airports. That's because Jetset uses the GPS data from your phone to figure out where in the world you really are. If your actual GPS coordinates match any of the hundred airports in the game's database, you get access to a customized airport game level that perfectly matches your real-world location. Complete that level, and you unlock a local achievement, or, in Jetset-speak, a "souvenir." For example, at Albuquerque International Sunport you can earn a virtual green chili pepper, while at Los Angeles International Airport, you win giant virtual sunglasses.

Every time you earn a souvenir, you can use the game's mobile Facebook application to send the virtual object as a gift to a friend or family member.

The gift lets them know not only that you've scored a game victory at the airport, but also clues them in to the fact that you're just about to start or finish a trip. In other words, Jetset helps you provide real-time travel updates to your social network as you play.

The more airports you visit, the more strange items you can amass for your souvenir collection and the more travel trophies you can collect. And if you're always flying in and out of the same airports? Then you can work on harder and harder levels to earn premium versions of your local souvenirs. Fly often enough in real life, and you'll get promoted up the virtual security ranks at your local airport. It's essentially FarmVille for airports, providing players with a sense of blissful productivity and social connectivity in an otherwise stressful environment. And that's what makes Jetset an alternate reality game, and not just another diversion. It's meant to improve players' real-life experience of a real-world environment.

Do the virtual souvenirs and power-ups have real value for the players? Bogost certainly hopes so. He specifically designed them to give frequent fliers something more fun and personally satisfying to aim for than miles and upgrades.

"Too many business travelers are obsessed with getting more miles even as they complain about how much they travel," Bogost told me. "It's a self-defeating system: it rewards you with more of what you already hate." Not to mention, relying on rewards of significant monetary value to keep people happy and motivated simply isn't a scalable solution. There's only so many free seats airlines are willing to give away, and only so many VIP members they're willing to recognize. As soon as too many people start earning rewards, Bogost notes, airlines simply change the rules to make it harder to win. That's not a very fair game.

By contrast, the potential intrinsic rewards of a good game like Jetset are nonexhaustible. Positive emotions can be provoked for everyone who plays, without limitation, and personal feelings of satisfaction, pride, and social connection are completely renewable resources. You can simply reward more people more often when the goal is an intrinsic reward.

Nothing epitomizes mandatory, mindless activity more than waiting in line at the security or boarding lines at the airport. But Jetset is a special, *voluntary* mission you can undertake while waiting—in other words, an unnecessary obstacle. By focusing on the unnecessary obstacle of the game, instead of the mandatory obstacle of the real security and boarding process, you can instantly change your state of mind from negative stress to positive activation. You can't opt *out* of security and boarding rituals, but you can opt *in* to the game. It's a subtle, but powerful, way to change the dynamics of the situation. Instead of feeling external pressure, you're focused on the positive stress of the game.

What I like about Jetset most is the fact that when you play, you're not just sleep-walking through a part of your life that you hate. You're actively participating in the moment, taking full advantage of your location by undertaking a game mission you could *only* play while at that airport.

Taking full advantage of the moment is an important quality-of-life skill: it builds up your sense of self-efficacy by reminding you that you have the power at any time to make your own happiness. Jetset might not permanently resolve the ongoing frustrations of airport security and boarding, but it reminds us of our power to improve our own experience. And for that reason, it's an excellent signal of the role that *location-based games* can play in improving our quality of life in the future.

A good location-based game can transform any space into sites of intrinsic reward. Imagine the possibilities. Three of my favorite potential game sites are dentist offices, the department of motor vehicles, and public transportation.

Wherever there is a mandatory experience that is unpleasant or frustrating, a surefire way to improve it is to design a good game you can *only* play in that space. Jetset effectively tackles that problem for airports. But what about the experience of actually being in the air?

Enter the Day in the Cloud challenge.

Accept the challenge.
Scour the earth.
Please remain seated.

—Invitation to play Day in the Cloud[4]

Take two ordinary commercial flights, flying at the same time in opposite directions between the same two airports. Pit them against each other in an epic battle of online wits and creativity. Passengers spend the duration of the flight working together to earn as many points for their plane as possible. When both planes land, everyone on the plane with the highest score wins.

Day in the Cloud was a promotion dreamed up by Virgin America and Google Apps. It was initially run as a small playtest, on planes traveling between the Los Angeles and San Francisco international airports. And while it hasn't yet been implemented across the Virgin America fleet, it serves as a powerful indication of the kind of innovation that is possible in the air, using technology that's already fully in place.

The game takes advantage of Virgin America's sophisticated in-flight entertainment system, which includes seat-to-seat chat and instant messaging; a real-time Google map that displays the plane's location, altitude, and speed; and WiFi Internet access for laptops, mobile phones, and PDAs.

Once the plane gets above ten thousand feet—which is when the plane's WiFi system is turned on—players can power up, log in, and join the game, which consists of a series of several dozen puzzles and creative challenges that must be completed before the plane descends back below ten thousand feet.

Each puzzle and challenge corresponds to a different altitude—the higher the altitude, the trickier it is. A low-altitude puzzle, for example, might be as simple as completing a maze or answering a movie trivia riddle, such as: "*Ma'am, I believe you are doing more than just flirting with me*. What 1967 movie features a more memorable version of that line?" (Check the footnotes for the answer.)[5]

Higher-altitude puzzles involve trickier tasks, like Mensa-level code breaking, and juicier goals, like snooping through a game character's "real" online e-mail account to find secret bits of personal information. And if you're not a puzzle person, you can tackle creative challenges, like: "Write a theme song for Day in the Cloud. The lyrics should have one four-line verse and one catchy four-line chorus. You must include at least one rhyme for 'cloud,' 'cirrus,' 'stratus,' 'cumulus,' or 'nimbus' somewhere in your lyrics."[6]

The collection of puzzles and challenges is designed to be virtually impossible to complete alone over the course of the flight. That's where your planemates come in. ("Planemates" might not be a recognized English word yet, but that's simply because we've been woefully underutilizing planes as social spaces.) Travelers are encouraged to work together, dividing and conquering the various challenges, and sharing solutions. You can partner with someone in your row, sharing a laptop together. Or you can use the seat-back communications system to trade ideas and answers.

The more passengers who play on a given plane, the higher the plane's potential score. So there's a real incentive to reach out to people who look friendly, curious, or just plain bored. And it's not just planemates that you can collaborate with during the game. The online game requires players to connect to the Internet, and once you're online it's easy to pick your friends' and family members' brains via e-mail or Twitter or IM. In fact, many Day in the Cloud players set up informal Twitter teams on the ground to help them out during the game. (Not everyone on the chosen flights knew about the game in advance—but one of the game's organizers told me afterward that about a dozen people on board each flight came prepared to play.) These on-the-ground collaborators serve as a kind of personal support system during the flight—not only good for the game, but also good for any anxiety and boredom you might ordinarily feel while flying.

A game timer shows you how long you have left in the flight, which is how long you have to finish solving your puzzles and completing your creative challenges. After the plane descends below ten thousand feet, the final scores are calculated and reported to both planes. As one player blogged after the flight, "Suddenly, I hear a big cheer come up from the whole plane behind me. 'We've won!'"[7] Winning passengers are greeted by Virgin America gate agents like conquering heroes when they walk off the plane.

All in all, it makes for quite a brilliant image: two planes passing in the sky, one heading north, the other south, trying to solve the same problems from above the clouds as they race along at hundreds of miles an hour.

Okay. So maybe this sounds fun, but you're still thinking: Why bother? Why add games to flights, when they already do what they're supposed to—get us safely from one part of the world to another? Do we really need to have "fun" and "adventure" and make "progress" all the time?

No, of course we don't.

If you're a good sleeper or worker on flights, or the kind of person who can relax and read a good book or just enjoy the view, then tuning out the game would be easy. You can and should go about your travel reality as usual. Many people will—during the Day in the Cloud playtests, roughly half the fliers on the test flights chose to play, while the other half went about their business.

But flying is difficult for many millions of people. It causes untold stress, anxiety, exhaustion. When something is that hard for so many people, when it causes so much daily suffering, needlessly, we should try to make it better if we can. If you're a nervous flier or get bored easily or just can't sleep on planes, an in-flight game could provide the kind of engagement and positive stimulation you need to actually start to enjoy and appreciate flying.

Day in the Cloud demonstrates quite clearly that the technology and desire is already here for a very different travel reality.

Consider some other possibilities. For example, an in-flight-only role-playing game that remembers exactly where you left off and picks up again whenever you board the plane. Its fantasy maps would overlay perfectly on top of the real-time Google maps. Each quest could be undertaken only while you're actually flying through that part of the realm.

Collaborative, GPS-enabled challenges would require you to partner up with someone on the ground within a hundred-mile radius of your plane and synchronize your virtual actions together as you fly overhead. Suddenly, flying over Nebraska is very different from flying over Kansas—because perhaps you have allies in Nebraska who can help you score more points, if you can get them to log on and play during those exact fifteen minutes you're flying overhead.

Of course, frequent-flier miles could also be made to be much more useful than they are now. For instance, you could distribute them as experience points across various categories of skill, talent, and ability to power up your in-flight avatar.

In-flight games even suggest a new model for earning seat upgrades—first player to score a certain number of points wins a first-class seat. As one Day in the Cloud player reported from the playtest, "At this point one of the attendants asks if I would like to move to first class since there's more room and I'm effectively the star player. I'm a bit reluctant, being that I'd lose my newfound friends sitting next to me."[8] (In case you're wondering, he eventually convinced the attendant that they should all move up together, so they could keep collaborating.)

Ultimately, when every mile you cover in the air is a chance to rack up more mission points, and every passenger on the plane is a potential ally, and flying over a town or city is a chance to connect with the people who live there, the whole experience becomes charged with potential to do more than just get where you're going.

THE EXAMPLE OF in-flight games presents the basic case for developing games that connect with our everyday lives: these games can help people suffer less and enjoy the real world more. When an experience is difficult for us, offering challenging goals, tracking points and levels and achievements, and providing virtual rewards can make it easier to get through the experience. Ultimately, that's the most important work that game designers can do in the future: to make things that are hard for us as rewarding—as *intrinsically* rewarding—as possible.

But what about activities that we already enjoy?

Can games motivate us to make a better effort, even when we already love what we're doing?

Trying to improve an already enjoyable activity by adding points, levels, and achievements has its risks. Economists have demonstrated that offering people an extrinsic reward (like money or prizes) for something they're already doing—and already enjoying—actually makes them feel *less* motivated and *less* rewarded. But game points and achievements don't have extrinsic value yet—so as long as the main prize is glory, bragging rights, and personal fiero, the danger of devaluing a pleasurable experience with game feedback is relatively low. But it's not nonexistent. Like money or prizes, the opportunity to earn points and level up could potentially distract us from the initial reasons we like to do an activity.

Clearly, we have to be thoughtful about where and when we apply game-like feedback systems. If *everything* in life becomes about tackling harder challenges, scoring more points, and reaching higher levels, we run the risk of becoming too focused on the gratifications of positive feedback. And the last thing we want is to lose our ability to enjoy an activity for its own sake.

So why risk it at all? Because measuring our efforts with gamelike feedback systems makes it easier for us to get better at any effort we undertake. As the great nineteenth-century mathematical physicist Lord Kelvin famously said, "If you cannot measure it, you cannot improve it." We need real-time data to understand our performance: are we getting better or worse? And we can use quantitative benchmarks—specific, numerical goals we want to achieve—to focus our efforts and motivate us to try harder.

Real-time data and quantitative benchmarks are the reason why gamers get consistently better at virtually any game they play: their performance is consistently measured and reflected back to them, with advancing progress bars, points, levels, and achievements. It's easy for players to see exactly how and when they're making progress. This kind of instantaneous, positive feedback drives players to try harder and to succeed at more difficult challenges.

That's why it's worth considering making things we already love more gamelike. It can make us better at them, and help us set our sights higher.

Nike+

Let's consider the gamelike Nike+ (or "Nike plus") running system, a motivational platform that is wildly popular among people who already love to run—especially those who want to run farther and faster.

> *Nike+*
> Stats! Stats! It got me out of bed to run this morning cuz I need BETTER STATS. It's real world achievement points! Who else will play with me? I seek challengers![9]
>> —Message board post from a new Nike+ runner

The very first time I went running with the Nike+ system, I ran faster than I had in my entire life.

I was running my favorite route, a four-and-a-half-mile course in the Berkeley Hills. In six years, running it a couple times a week, I'd never once finished faster than 41:43. But on my first Nike+ run, I clocked in at 39:33, more than two minutes ahead of my all-time personal best. How in the world did I suddenly get so much faster? It's no mystery: I was motivated by better, real-time feedback and by the promise of online rewards when I got home.

Running, of course, is its own reward. You feel the endorphins, you clear your mind, you build stamina, you burn calories, you get stronger. But it's also a struggle—to find the time, to convince yourself that you have the energy when you'd rather sleep late, to go out whether it's hot or it's raining, and to fight off boredom doing a highly monotonous activity. Runners love running, but motivation is still an issue. So Nike+ is designed to provide an added layer of intrinsic motivation, beyond the runner's high and the physical results.

If you've never seen it in action, here's how Nike+ works. An inexpensive sensor—it costs about twenty dollars and is smaller than a poker chip—fits imperceptibly inside the sole of almost any standard Nike sneaker. It's activated by movement (thanks to an accelerometer) and communicates with your iPod (via radio transmitter) to tell you exactly how fast you're running and how far you've run. As you're running, presumably to your favorite music, the iPod screen displays your stats in real time.

Getting feedback in real time makes a huge difference when it comes to running faster and longer. Just being able to see when you're slowing down—something that happens unconsciously as you tire or lose focus—helps you bring your attention back to your pace. Meanwhile, pushing yourself to run faster is instantly more rewarding, because you get to see the numbers drop lower and lower the faster you go. It's one thing to set a time goal and try to reach it; it's another thing entirely to know every step of the way if you're running fast enough to achieve it.

When you get home, you can plug your iPod into your computer, and the Nike+ system will upload your data and add it to your running profile. That's where the online rewards come in. Every mile you run earns you a point; score enough points, and you level up. There are six levels currently on Nike+, which follow the same color grading as martial arts belts: yellow, orange, and green; blue, purple, and black. Like any good MMO, you advance Nike+ levels quickly at first, but over time it takes more and more effort to reach the next level. Right now, I'm a level green runner, having logged 272 miles since joining, and I have 348 more miles to run to reach the blue level. That's an intimidating number, but I'm so motivated to level up that I bet I'll run the next 348 miles in even *less* time than I ran the first 272.

Based on the data the Nike+ sensor collects, you can also earn personal on-line trophies for best times and longest runs, as well as achievements for meeting training goals, like working your way up to a 10K distance or running a hundred miles in a hundred days. And when you've had a particularly good run, a famous athlete like Lance Armstrong will cheer you on before you even catch your breath, with a congratulatory audio message like this: "Congratulations! You've just recorded a personal best for the mile" or "Way to go! That was your longest run yet."

You can keep your running profile private and your accomplishments to yourself—if you want. Or you can push your stats and achievements out to your Nike+ friends online, to everyone you know on Facebook, or even to the whole world on Twitter. Perhaps my favorite Nike+ motivational feature is the "power song." It's the musical equivalent of a health pack or a power-up in a video game. Whenever you need a boost of energy or extra motivation to keep running or pick up speed, you simply hold down the center button on your iPod. That quick gesture automatically triggers your favorite, preset running song. For me, pressing the center button during a hard run feels like I'm unlocking some secret super-running power that I didn't even know I had. The faster pace, the pounding beat, the lyrics ringing in my ears like a personal mantra—it's the one time in the real world I feel like I have the ability to summon the kind of magical powers that I'm used to deploying in virtual worlds.

Add all that up—real-time stats, a leveling system, personal achievements, and your own personal power-up song—and Nike+ makes for a very good running game, one that uses better feedback and reward to help you put in a better effort and aspire to more than you would otherwise. But why play alone when you can play with others? It's the online community built around the Nike+ system that turns it into something really spectacular: not just a running game, but a massively multiplayer running game.

The Nike+ online community has more than 2 million active members, all of whom are collecting and sharing data about their runs in order to compete in challenges and contribute to team missions.

Anyone can design their own challenge and invite whomever they want to play with them. It can be competitive—everyone tries to get the best score—or collaborative—you try to get all of the participants to successfully finish the challenge before time runs out. Challenges can be as small as a two-player rivalry—husband versus wife or brother versus brother, for example: Who can log the most miles in a week? Or they can be set up as a team event for a group of friends or coworkers, with a dozen, or twenty, or fifty runners, or more—one neighborhood races another, for example, or every department for itself: how many teams can collectively log a thousand kilometers before time runs out?

The challenges can also be public free-for-alls, with hundreds, thousands, or even tens of thousands of competitors. As I'm writing this, there are more than seven thousand user-created public challenges to participate in, including the collaborative individual challenge of "running around the earth," in which each participant runs 24,902 miles—the challenge expires in the year 2027, making this ambitious goal seem a bit more reasonable—and a competitive team challenge for runners who go out with their dogs. (In this public challenge, players can join a team based on breed; out of fifty different teams, currently Labradors and beagles are leading the total mile count, followed closely by the mutts, but the Australian shepherds have the fastest pace.)

The challenge puts the runner's personal goals into a larger social context, which gives each jog more meaning. Every run is adding up to something—and depending on what motivates me most, I can join challenges that stoke my competitive spirit or call on my sense of responsibility to my teammates.

My Nike+ Mini trash-talks me.
(Nike Corporation, 2009)

Of course, no good MMO would be complete without an avatar. Nike+ is no exception. When you join the Nike+ community, you get to create a "Mini," officially described as your "tiny running partner," whom you can customize to look just like you. Your avatar's energy level and animations are based on your run activity: how far and how often you run. If you've put in a few good days in a row, your Mini is ecstatic and bouncing off the walls. If you've slacked off for a week or two, your Mini pouts and mopes and gently teases you for being such a slacker. Just a few days ago, my Mini was making faces at me and saying, "If only I practiced running like I practice paddleball."

Your Mini greets you whenever you log in to Nike+, you can embed it into your Facebook profile or blog (so others can see your avatar), and you can even download a screen saver starring your Mini at play (so you have to come face-to-face with your avatar even when you're not thinking about running).

Recent research suggests that this kind of ambient avatar feedback is remarkably effective. In a widely cited experiment conducted at Stanford University's Virtual Human Interaction Lab (VHIL), researchers demonstrated that watching customized, look-alike avatars lose or gain weight as we do exercise makes us work out longer and harder.[10] Participants who received "vicarious reinforcement" from their avatars volunteered to do on average eight times more exercise repetitions than participants without avatar feedback. That bodes well for the potential use of Mini-like avatars at home or at gyms, where people are more likely to work out in front of screens. (And, in fact, many home fitness games, including Wii Fit and EA Sports Active, use avatar feedback to engage players in harder workouts.)

But there's no reason that people working out need to be stuck in front of a screen to get the benefit of avatar feedback. In another experiment, Stanford VHIL researchers discovered that simply showing subjects a short animation of their look-alike avatar running in the laboratory inspired subjects to spend on average an hour more running in the first twenty-four hours after they left the laboratory. (There was no motivation effect watching a random avatar; it worked only when the avatar was highly customized to look like the subject.)

The researchers theorized that seeing virtual versions of themselves doing a positive activity stimulated memories of the subjects' own real-life positive experiences, making them more likely to reengage in the activity. They were careful to note in their findings that participants in the study, all college-age students in northern California, were generally healthy and fit. There was no evidence to suggest that someone who hates running would be likely to run for an hour after seeing their avatar do it. The avatar reinforced positive feelings about running, rather than creating them from scratch.

Yesterday, after my first run in a couple of weeks, my Mini danced around my iPod smiling, saying, "I can hardly contain myself! I'm a running machine!" Today, after another run, she's leaping over hurdles and shouting, "I can do anything! I feel amazing!" I have to admit—the animations are a fairly accurate depiction of my own inner runner. It's definitely working the way the Stanford researchers theorized it should: my Mini reminds me of why I love to run and therefore makes me more likely to get out of the house and do it.

But there's also something else going on. I find that I want to run more in order to make the Mini happy.

Though it might seem ridiculous, this kind of emotional connection happens in games all the time—especially in tending and caretaking games, like the Xbox Viva Piñata series, in which players have to support an ecosystem of "living," wild-roaming piñata animals, or the Nintendo *Pikmin* series, which puts the player in charge of an army of eager-to-please but dumb and highly vulnerable creatures. MIT researcher Judith Donath has studied the emotional attachment we form to virtual creatures. She argues that game characters programmed to appear dependent on us for their well-being provoke a hardwired human desire to nurture and care for them, and it doesn't hurt that they are cute, helpless creatures. "Time spent playing with them feels like care-taking, an act of responsibility and altruism," Donath explains. "We develop empathy for them and become invested in their well-being."[11] Naturally, then, the happier our virtual creatures appear to be as a direct result of our actions, the more satisfied we are as effective caretakers.

Virtual-creature happiness is not nearly as obvious a feedback system as points, levels, and achievements. But it's part of a larger potential field of reward innovation, as we continue to learn how to better motivate ourselves by applying the best design strategies of games to our real-life activities.

THE MORE we start to monitor and self-report our daily activity, whether through GPS, motion sensors, biometric devices (to track heart rate or blood sugar levels, for example), or even just with manually entered status updates, the more we'll be able to chart our progress, set goals, accept challenges, and support each other in our real lives in the way we do in our best games. Given the overwhelming success of the Nike+ system, it's not difficult to imagine adopting some of the Nike+ strategies for other activities that we want to do faster, more often, or simply at a higher level.

I for one would have loved a Writing+ system while writing this book. I'd have a "mini" writer whose mood and energy was based on my daily word count. I'd have the opportunity to earn achievements, like showing up to write ten days in a row, or to set a personal best for most words written in a day. The system could also keep track of the complexity of my writing—how many words I use per sentence, and how many sentences per paragraph, for example. I could use this data to improve the clarity of my writing and vary its structure. I could set up friendly rivalries with other authors—both friends in real life and authors that I'm a fan of. I think I would have been a lot more inspired to write if I knew I'd be able to compare my daily writing stats against the real-time stats of my favorite fiction writers—Curtis Sittenfeld, Scott Westerfeld, Cory Doctorow, and Emily Giffin.

Any project or challenging hobby that we're working on that we want to see through to completion would benefit from more gamelike feedback and ambient support. We may be looking at a future in which everything we do can be "plus": Cooking+, Reading+, Music+.

Maybe even . . . Social Life+?

That's the idea behind Foursquare, a social networking application designed to motivate players to lead a more interesting social life.

Foursquare

The premise of Foursquare is simple: you'll be happier if you get out of the house more and spend more time face-to-face with your friends.

Created by independent New York City–based developers Dennis Crowley and Naveen Selvadurai, Foursquare takes its name from the classic red-rubber-ball playground game. To participate in Foursquare, you simply log in to the mobile phone application whenever you show up somewhere public that you deem fun, then tell the system where you are. That's called a "check-in," and you might find yourself checking in from a restaurant, bar, café, music venue, museum, or wherever else you like to go. Whenever you check in, Foursquare then sends real-time alerts to your friends so they can join you if they're free and in the neighborhood. It also lets you know if any of your friends are already nearby, in case you want to meet up with them. Most importantly, Foursquare keeps track of where you've been, when, and who you've checked in with, if they're playing Foursquare, too. By mid-2010, more than a million people were tracking and sharing data about their social lives using the Foursquare system. And more than three-quarters of those users were checking in thirty or more times each month.[12]

Out of all this data, Foursquare produces a series of online metrics about your social life: how often you go out, how many different places you visit, how many different people you spend time with each week, and how frequently you visit your favorite spots. On their own, these metrics aren't that interesting. They're just data, a way to quantify what you're already doing. What really makes Foursquare engaging is the challenge and reward system built around the data.

The most popular Foursquare feature is a competitive challenge called The Mayor. The rules read: "If you've got more check-ins than anyone else at a particular place, we deem you 'The Mayor' of that place. But once someone else comes along who has checked in more times than you, they then steal the 'Mayor' title back from you." As soon as you become mayor, Foursquare sends an announcement to your friends congratulating you. Even better, some bars and restaurants have set up special deals for whoever happens to be mayor at any given time. The Marsh Café in San Francisco, for example, lets the current mayor drink for free. Of course, this is also a smart move on the part of the café—players have extra incentive to bring their friends there nightly to try to achieve or hold on to the mayor status, boosting business throughout the week. It's also a good example of how traditional brick-and-mortar companies might be able to augment their services by more actively taking part in this popular reality-based game. Currently, hundreds of venues—from the Sacramento Zoo to a Wendy's fast-food restaurant in the student union at the University of North Carolina Charlotte—offer deals or freebies for Foursquare players.

Why do people love the idea of becoming the mayor? Because trying to become mayor of your favorite city spots gives you a chance to keep doing something you already love, but do it more. It gives you an excuse to spend as much time as possible at the places that make you happiest. And when you notice someone else vying for your mayor status, you get an instant friendly rival, motivating you to visit your favorite places more often, the same way a Nike+ challenger pushes you to run faster and longer.

Foursquare is also a personal achievement system, consisting of virtual trophies and badges. Trophies automatically unlock in your profile when you celebrate checking in to your tenth, twenty-fifth, fiftieth, and hundredth different venues in a single city. In order to earn these trophies, you can't just be content with being the mayor at one place. You have to venture outside your usual spots and expand your social horizons. You can also earn badges like the Foodie badge, earned by checking in to Zagat-rated restaurants in New York, San Francisco, Chicago, and other major cities, or the Entourage badge for checking in at the same time and place as ten or more of your Foursquare friends.

In the end, what makes a Foursquare social life better than your regular social life is the simple fact that to do well in Foursquare, you have to enjoy yourself more. You have to frequent your favorite places more often, try things you've never tried before, go places you've never been, and meet up more often with friends whom you might not ordinarily make time to see in person. In other words, it's not a game that rewards you for what you're already doing. It's a game that rewards you for doing new things, and making a better effort to be social.

There's one more significant benefit to adding compelling stats to your social life. Because players want their statistics to be as accurate (and impressive) as possible, they're more likely to remember to check in and send status updates about where they are. That makes it easier for their friends to find them, and therefore more likely to make plans to see them.

Ultimately, the real reward of seeing friends more often and breaking outside your routine has nothing to do with virtual badges or social life points or online bragging rights. The real rewards are all the positive emotions you are feeling, like discovery and adventure; the new experiences you're having, like hearing more live music and tasting more interesting food; and the social connections you're strengthening by being around people you like more often. Foursquare doesn't replace these rewards. Instead, it draws your attention to them.

Some people, of course, are natural social butterflies or nightlife adventurers. For others—workaholics, homebodies, introverts—getting out and doing something new is no small feat, especially when there are so many compelling reasons to stay in our own living rooms.

There's a popular gamer T-shirt that shows an Xbox Live–style badge of a door ajar with these words alongside: "Achievement unlocked: Left the house."[13] It's a joke, but it also speaks to the real challenges of trying to lead a meaningful, balanced life in the nonvirtual world. As we struggle to find the right balance between virtual and real-life adventures, a game like Foursquare can nudge us in the right direction and help us put our best efforts where we can reap the most satisfying rewards: back in the real world, with the help of a good game.

CHAPTER NINE

Fun with Strangers

HOW ALTERNATE REALITY GAMES CAN CREATE
NEW REAL-WORLD COMMUNITIES

It's a cold and dreary afternoon, and you're walking down a busy street. You're lost in your thoughts when suddenly a woman's voice whispers in your ear, "There's a lover nearby . . ." You look around, but everyone seems as lost in their own world as you were just a few seconds ago. If there's a lover nearby, you have no idea who it is.

Then you hear the voice again, this time updating you on your game statistics: "Your life is now at level six." That's one level higher than it was before the lover passed by.

Some stranger on the street just gave you a life.

But who was it? Is it that kid sitting on the steps now a few buildings behind you, with his earbuds tucked in? He looks like he's listening to music—but is he listening for lovers, too? Or is the lover that man in the suit with his Bluetooth earpiece, pacing back and forth? He looks like he's on a business call—but could he be your secret benefactor?

Or has the lover moved on? Perhaps you are on your own again.

You haven't gone another half block when the voice interrupts, this time more insistently, "There's a dancer nearby." Then, right away: "There's an-

other dancer nearby. Your life is now at level four." Damn! Who just stole two lives from you?

It must be a couple, playing together, because you've lost two lives in such rapid succession. You spin around and notice a couple holding hands walking in the opposite direction. They might be wearing headphones under their hoods. You didn't notice them before, but they must be the dancers. You hurry down the block before they circle back and take another life from you.

Clearly, you need to find some other lovers as quickly as possible, team up, and restore each other's life levels. If your life falls to zero, you're out of the game. But how do you discover the other players hidden in the crowd? As the game instructions suggest, "You could find a stranger and ask them, 'Are you a Lover or a Dancer?'" But that feels too forward, too abrasive. You feel more comfortable scanning the crowd, looking for people who seem to be looking for others. That way, you can gravitate toward the most promising strangers, stand near them, and wait to see if your life level goes up or down.

If nothing happens, you know they're not playing the game and you don't have to bother them. But if your life level goes up, you can try to smile and make eye contact. You can try to show the stranger that you can be trusted. . . .

Learning how to offer comfort to strangers, and how to receive it, is the primary challenge of a game called, naturally, the Comfort of Strangers. It's a game for outdoor city spaces, designed by British developers Simon Evans and Simon Johnson. It's played on PDAs and phones with Bluetooth detection that alert you via your headphones or earpiece whenever other players are within a few yards' distance. The PDAs automatically detect other players within a few yards and register a gain or loss of life whenever you cross paths. Half the players are "lovers"; they form one team. The other half are "dancers," and they form the opposing team. If you encounter a player on your team, you gain a life; if you encounter a player on the opposing team, you lose one.

The Comfort of Strangers is played anonymously; you can download and start the application and wander out into the city streets without any idea of who else is playing or how many players there are. There's no visual or screen element to the game, so you can play it quite discreetly, with your PDA tucked into a pocket. The only clue that you're playing is that you're wearing head-phones—but it's easy to blend in with the increasing number of people who wear earbuds or earpieces while out in public spaces.

At the start of the game, you don't know what side you're on. You have to learn whether you're a lover or a dancer by listening to the voice that whispers in your ear and keeping track of whether your life is going up or down. Every-one starts the game with ten lives, and when only one team remains alive, the game ends.

According to Evans and Johnson, the game is designed to evoke the feelings of loneliness and anonymity that are a mainstay of urban life—as well as to provide opportunities for strangers to mean something to each other, if only briefly. As they explain, "The game immerses players in the crowd, exposing them to the ambivalent feelings aroused by city life, the freedom of anonym-ity and its loneliness. Out of the drive to stay in the game, players create ad hoc, or improvised, social groups."[1] They have to develop their intuition about how to tell who else is playing, and therefore who represents a part of the game community. They learn to see strangers for the potential relationships they represent, not just as obstacles to avoid as they pass by.

The emotional impact of the Comfort of Strangers is intense. It not only heightens your awareness of the potential for strangers to play a role in your life, it also provokes a real curiosity about others, and a longing to connect. When you start the game, you feel like you might be the only one playing. Each time you encounter another player, it's reassuring—even if they're on the other team. When I asked Simon Johnson about the social goals of the game, he told me this was intentional:

> We wanted our players to find some way to connect with the strangers around them, so we tried to make them feel lost and alone. We set the game up to create a degree of uncertainty in players as to who was and was not playing. We played with the boundary between players and nonplayers so that finding another playing stranger always brings you comfort, even if they're on the

opposite side. Because at least they understand your actions, they
understand that you are part of the same game.[2]

The Comfort of Strangers can be a short game or a long game, depending
on how willing players are to overcome their hesitations about reaching out
to strangers, and depending on how tightly they can learn to stick together in
the crowd.

In theory, if such a game became immensely popular, you could play it all
the time, as part of your regular routine—you'd simply turn the game on whenever you walked outside and always keep open the possibility of running across
another player as you went about your ordinary business. But in practice, while
games like this are still relatively new, there isn't a critical mass of players to
accommodate continuous play. Instead, players organize games online and set
precise windows of time and playing fields: for example, in a certain neighborhood, during a certain hour, on a particular date. This kind of advanced schedule keeps players anonymous, but ensures there will be enough density of play
for players to have a good chance of encountering each other.

Because a critical mass is so important to games like the Comfort of Strangers, in 2008 Evans and Johnson cofounded an annual Bristol-based festival
called Interesting Games, or Igfest, for innovative outdoor games. The festival
is meant to provide support for and exposure to other game developers who
are working to make cities more interesting and friendlier spaces. And it's
one of an increasing number of urban game festivals worldwide—from the
annual Come Out & Play festival in New York City, founded in 2006, and the
Hide & Seek Weekender festival in London, founded in 2007, to the Urban
Play festival in Seoul, South Korea, founded in 2005—that are designed
to test the power of games to improve the feeling of community in real-
world spaces.

These outdoor game festivals gather critical masses of players together for
an entire week or weekend of games with the aim of helping to introduce
these games to the public at large. They also embody our ninth fix for reality
in action:

○ FIX #9: MORE FUN WITH STRANGERS

Compared with games, reality is lonely and isolating. Games help us band together and create powerful communities from scratch.

What does it mean to create a community from scratch?

It's hard to pin down the difference between a community and a crowd, but we know it when we feel it. Community feels *good*. It feels like belonging, fitting in, and actively caring about something together. Community typically arises when a group of people who have a common interest start to interact with each other in order to further that interest. It requires **positive participation** from everyone in the group.

In order to turn a group of strangers into a community, you have to follow two basic steps: first, cultivate a shared interest among strangers, and, second, give them the opportunity and means to interact with each other around that interest.

That's exactly what a good multiplayer game does best. It focuses the attention of a group of people on a common goal, even if they think they have nothing in common with each other. And it gives them the means and motivation to pursue that goal, even if they had no intention of interacting with each other previously.

Does a game community among strangers last? Not always. Sometimes it lasts only as long as the game itself. The players might never see or talk to each other again. And that's perfectly okay. We often tend to think of communities as best when they're long-term and stable, and certainly the strength of a community can grow over time. But communities can also confer real benefits even when they last for mere days, hours, or even minutes.

When we have community, we feel what anthropologists call "communitas," or spirit of community.[3] Communitas is a powerful sense of togetherness, solidarity, and social connection. And it protects against loneliness and alienation.

Even a small taste of communitas can be enough to bring us back to the social world if we feel isolated from it, or to renew our commitment to participating actively and positively in the lives of people around us. Experiencing a short burst of community in a space that previously felt uninviting or simply uninteresting can also permanently change our relationship to that space. It becomes a space for us to act and to be of service, not just to pass through or observe.

Comfort of Strangers designers Evans and Johnson believe that experiencing communitas in an everyday game can spark a taste for the kinds of community action that make the world a better place. Learning to improvise with strangers toward a shared goal teaches players what they call "swarm intelligence"— intelligence that makes people better able and more likely to band together toward positive ends. "As we're making these games, we dream of the other revolutionary things swarm intelligence might make possible. Low-carbon futures, mass creativity, living happily with less."

It's not such a radical idea. To see why, let's look at two other games designed to create unexpected moments of communitas in a specific shared space: Ghosts of a Chance, a game for a national museum, and Bounce, a game for a retirement center. Both groundbreaking projects demonstrate the growing importance of having more fun with strangers and of using games to build our own capacity for community participation.

Ghosts of a Chance: A Game to Reinvent Membership

Most museums offer memberships where members pay an annual fee and can then visit the museum as often as they'd like. It's a good way to raise money and promote visitation, but it's not a particularly good way to experience membership. Members of the museum are, for the most part, like any other visitor: they take in the museum's offerings, but don't interact with other members, or even know who they are.

Recently, the Smithsonian American Art Museum set out to experiment with a new model of museum membership, a way to *really* belong to a museum. It's a model that calls for members to contribute real content to the museum's collection and to collaborate with each other online in between museum visits. To test this more participatory model of membership, the Smithsonian developed a six-week alternate reality game called Ghosts of a Chance for one of its main facilities, the Luce Foundation Center for American Art.

The Luce Foundation Center is described as a "visible storage facility" for the Smithsonian. It displays more than thirty-five hundred pieces of American art, including sculptures, paintings, craft objects, and folk works, in densely packed floor-to-ceiling glass cases. Its primary purpose is to display as much of the vast Smithsonian collection as possible, much more than can typically be included in the other galleries.

Because it's so packed with art, visiting the Luce Foundation Center is a bit of a treasure hunt already: among all the diverse pieces, you have to seek out the special objects that speak to you most. The center has at the core of its mission teaching visitors to really hear what the art objects have to say, and its educational materials often focus on how art is a window into the lives and times of its creators. There's a sense in the museum that history lingers in the art objects almost like a ghost, waiting to whisper its tales to visitors. Learning how to hear those tales, and how to whisper our own histories through artwork, was the inspiration for the Ghosts of a Chance game.

The game begins with what at first seems like a real press release from the museum. Members, as well as public visitors to the museum's website, are invited to meet two new curators at the center, Daniel Libbe and Daisy Fortunis. According to the press release, they will both be writing about their work on blogs and their social network pages. Read the fine print, however, and you realize Daniel and Daisy aren't real curators. They're fictional characters in a new, experimental game produced by the Smithsonian. And if you want to find out more, you have to friend these fictional characters on Facebook and start following their blogs.

That's when you discover that Daniel and Daisy are having a rather extraordinary experience: they're communicating regularly with two ghosts haunting the Luce gallery, a man and a woman who lived a century and a half ago. Angered at being forgotten by history, the ghosts are threatening to destroy the museum's precious artifacts—and they won't rest until *their* stories are represented in the museum's glass cases.

Frightened but resourceful, Daniel and Daisy make special arrangements for a one-day exhibit called, naturally, Ghosts of a Chance. But ethereal ghosts can't make real art—so Daniel and Daisy need the museum members to help. It's up to them, the players, to interpret the two ghosts' histories—by transforming their tales into art objects that the curators promise to display in a special gallery event.

And so a gameplay mechanism is established. Each week, the ghosts reveal a new dramatic chapter in their lives to Daniel and Daisy, describing in mysterious terms the kind of art piece that they feel would best capture their secret histories. Daniel and Daisy then pass on the new information to members of the game and charge them with the important mission of making that art real, then sending it to the Smithsonian for inclusion in the exhibit.

In the first mission, for example, players learn that one of the ghosts is tortured by memories of a dear friend, a young lady from a very wealthy family:

> She's a girl from another time, she blushes and rustles as she passes, taffeta skirt buoyed by crinolines. She has taught herself to fling her burnished curls with just a turn of her head; she and her sister practiced for hours in front of an oval mirror. At twenty, she is poised; she understands her value; her next great adventure awaits her. A mate. Travel. Then, domesticity—which involves a love of gardening, cleanliness and the proper care of servants. . . . [4]

Players are then challenged to craft this girl's most prized piece of jewelry, what the ghosts call the Necklace of the Subaltern Betrayer. Instructions for designing the necklace are spare, and poetic: "The Necklace I want should fit perfectly around her neck, but remain there only long enough for me to steal it right off again."

Players discussed the challenge in online forums: What does "subaltern" mean? (They learned that it is a political-science term for people who lack power or social status in a given society.) They debated: Should the necklace be old-fashioned, or a modern interpretation of the tale? They collaborated to unpack the meaning of the tale, to analyze the cultural clues embedded in it, and to strategize about how to craft a necklace that could evoke such a story and communicate such intense feelings.

As a community, the players decided the necklace should convey what it would feel like to wear the heavy and inflexible societal expectations of a woman of money and privilege. One player created a necklace titled "Someone to Watch Over Me," comprising more than a dozen squares of fabric, each screenprinted with the image of a different staring eye. The eyes are stacked on top of each other in geometric sets of one, two, and three, and strung along a pretty pink ribbon. The aesthetic is both girly and intimidating. Another player submitted a necklace titled "Enclosure," which appears to be constructed from barbed wire strung with rubies. Both the title and design of the work suggest that its wearer is trapped and limited by her social status, her riches preventing her from living the life she might otherwise pursue.

All of the player-created artifacts received by the museum were cataloged online and archived at the Luce Foundation Center. Players around the world could see the different interpretations of the challenge—either online or in person by visiting the objects on temporary display at the museum. In the end, more than six thousand Smithsonian members and fans participated in the online experience, while two hundred fifty attended the opening of the Ghosts of a Chance exhibit in person.[5]

Why design a game, instead of issuing an open invitation to design for the museum? There are two good reasons. Because it was a "game" and not a serious art competition, people who wouldn't normally feel capable of contributing artwork were free to try without risking embarrassment. The game structure, with its clues and narratives, also gave a larger and more atypical museum membership—in this case, mostly students and teenagers—an opportunity to participate in the making of the exhibit, through online discussion and analysis of the artworks, even if the members didn't contribute art themselves. These players helped serve as virtual "curators" for the Ghosts of a Chance exhibit.

To become a member of any community, you need to understand the goals of the community and the accepted strategies and practices for advancing those goals. Participating in Ghosts of a Chance educates museumgoers about both. Although it was a game, the participating players were treated seriously as both artists and curators. As Nina Simon, a leading expert on the use of technology in museums, reported at the time, "The game artifacts [were] officially entered into the collection database and stored (and accessed) the way other artifacts are—via appointment, white gloves, that sort of thing. In this way, the secret rules of museums become new hoops for the gamers to jump through—hoops that will likely add a level of delight as they expose the inner workings of the museum."[6] In other words, the gameplay knocks down the "fourth wall" that usually separates the work of the museum's curators and the visitors. And in doing so, it completely reinvents the idea of museum membership, making it possible for a real museum *community* to emerge.

We have become accustomed to viewing museums as spaces of consumption—of knowledge, of art, of ideas. Ghosts of a Chance shows how we can turn them into spaces of meaningful social participation, driven by the three fundamental components of gaming communities: collaboration, creation, and contribution toward a larger goal.

Bounce: A Game to Close the Generation Gap

What would it take to convince young people to call their grandparents more often? Better yet, what would it take to convince young people to call someone *else*'s grandparents while they're at it?

Those were the twin goals of a project called Bounce—a telephone conversation game designed to support cross-generational social interaction.

Bounce takes just ten minutes to play. When you call the game, you're connected live on the phone with a "senior experience agent"—someone at least twenty years older than yourself. You follow a series of computer prompts to swap stories about your past, in order to discover life experiences you have in common. For example: What is something you both have made with your own two hands? What is a useful skill that you were both taught by a parent? What is a faraway place you both have visited? Your goal: find out how many points of connection you can make with your senior experience agent before time runs out.

Bounce was the effort of a four-person team of computer scientists and artists at the University of California, myself included. We set out to design a computer game that would spark a stronger feeling of community across the generation gap.

There is a significant need for a game like this as retirement communities, senior centers, and continuous care homes can be very socially isolating. This is partly an environmental problem: they are typically single-use spaces, without significant cross-traffic, and there's little opportunity for the mingling of different age groups. But it's also partly a cultural problem: major studies at Harvard and Stanford have demonstrated that a prejudice against the elderly is one of the most widespread and intractable social biases, particularly in the United States and especially among people under the age of thirty.[7] Young people commonly associate older age with negative traits like diminished power, status, and ability, leading them to avoid interacting with people they perceive as elderly, even their own loved ones.

Our team spent the better part of a summer brainstorming potential concepts for a game to help bridge the generation gap more gamefully, and as part of that process I personally spent quite a lot of time on the phone with two important seniors in my life: my grandfather Herb, who was ninety-two years old that summer, and my husband's grandmother Bettie, who was eighty-seven. I was doing "user research" with them, figuring out what kind of game-play might be fun and easy to grasp quickly—particularly for older people who are not used to playing computer games—as well as to figure out the best way to get them to interact with the game technologies.

It was immensely rewarding to spend so much time on the phone with them. Phones were, of course, a familiar technology for both parties, easy for all concerned to access and use anytime. Talking on the phone was so rewarding and easy, in fact, that it eventually became obvious that giving young people a fun reason to call seniors on the phone should be the objective of our game.

But how do you build a computer game around making a phone call? We decided that we would make the game for two players at a time, since phone calls are most satisfying between two parties. The only rule? The players should be at least twenty years apart in age. Both players would need to be on the phone, of course, but since seniors are less likely to have constant access to a personal computer, we decided that only one of the two players should need to be in front of a computer. That player would log in to the game website, then call the other player.

We called the game Bounce, after the kind of exchanges we hoped to inspire: a quick, easy bouncing of life stories off each other.

The website prompts the players with collaborative interview questions: What's a body of water you've both swum in? What's a book you've both recommended to a friend? What's an experience that has made both of you nervous? The challenge is to discover a single answer for each question that is true for both players. Answers could include, for example, "We've both swum in the Pacific Ocean" or "We've both been nervous going on a first date." When you come up with a shared answer, one player types it into the game database. You have ten minutes to answer up to ten different questions total from the database of one hundred possible questions, and you can pass on any question. The game website counts down the ten minutes and reveals your score at the end.

We ran the game as an experiment for one week, based out of a senior recreation center in San Jose, California. With this kind of game, you don't want an open invitation to play; there needs to be a level of trust that the people calling will participate with a positive attitude. So we distributed the senior center's phone number via e-mail and social networks to trusted friends, family, and colleagues. We also gave out the phone number to attendees of an art and technology festival in nearby downtown San Jose, expecting that any-one who participated in the festival was more likely to be a positive player, and not a "griefer"—someone looking to spoil the game, rather than play it. We used a live matchmaker at the senior center to pair off the phone-in players with the seniors.

It was a bit of a risky proposition for both the seniors and the younger play-ers. After all, talking with a stranger can be awkward, but the game provided a clear structure and set topics for conversations. The fact that both parties on the call were working toward the same goal—a high score—created an instant connection.

The players' strategies varied. One kind of player would rush to list every place they'd been swimming, for example, while another would prompt their partner with inquiries like, "Where did you grow up? Were you near any lakes? Have you taken any trips to any oceans?" Common answers often prompted rushes of recognition, "Oh my god, wait—I do know that river! My parents took my sister and me rafting there when we were kids!" Uncommon answers just as often led to a chatty diversion from the gameplay: "You went swimming in Alaska? Did it hurt?"

Our prototype was highly successful. Nearly everyone who played once came back (or called back) to play again, and the senior players reported much higher moods after playing the game. The simple fact that they were described as "senior experience agents" in the game seemed to play an important role in their enjoyment. It set a playful tone and gave them confidence that they could participate. But perhaps the most successful design element was the score, which was both a number—your total answers out of a possible ten—and a poem.

We wanted both players to leave feeling like they had not only talked to each other, but created something together. So at the end of the game, the website turned the players' answers into a simple, free-verse poem. Players could print the poem out or e-mail it. Poems are also captured and viewable online. Here's an example of one of the free-verse poems that two players cre-ated as their final score together:

Rougemont, making wedding pictures, tango in a barn,
Bend paper clips, cinnamon buns, tongue of a cow,
In a skirt, in the Pacific, putting together a darkroom.

Each phrase of this poem represents something two strangers, at least twenty years apart in age, shared in common so far in their lives.

Now, I've never met the two players who both bend paper clips when they're nervous and have both tangoed in a barn, but I can imagine them, meeting each other for the first time on the phone and realizing how many shared events had led up to this moment in their lives.

In the end, the fun of the game is quite simple: the phone rings, and it's a stranger, and just by chance you get to discover someone very different from you who has nevertheless lived a similarly fascinating life. Of course, the game can also be played with relatives and neighbors, and more than once, because it can produce thousands of unique interview sessions.

When you start to realize how many interesting life experiences you might already share with someone from a completely different generation, there's no limit to the number of connections you can make. And a game like Bounce makes it much easier to reach out to someone whose life might benefit greatly from knowing you better.

THE THREE GAMES described in this chapter demonstrate how quickly and effectively a game can help us band together to experience a burst of communitas and participate more actively in the social commons.

Community games have important benefits to our real lives. They may lead us to new interests—public spaces or public institutions we discover we care about more than we'd thought, or activities like storytelling and art that we want to pursue with others. Even when the game ends, we may find ourselves participating more in these spaces, institutions, and activities than before.

On the other hand, the communitas we feel may be just a short spark of social connection, nothing more. But even playing a very short game together, we are reminded of how much we share with even the strangest of strangers. We gain confidence that we can connect with others when we want to, and when we need to.

And with that confidence, there is no reason to ever feel alone in the world—virtual, real, or otherwise.

CHAPTER TEN

Happiness Hacking

HOW ALTERNATE REALITIES CAN HELP US ADOPT
THE DAILY HABITS OF THE WORLD'S HAPPIEST PEOPLE

S hout compliments at strangers on the sidewalk.

Play poker in a cemetery.

Dance without moving your feet.

Maybe these aren't exactly your typical doctor's orders. In fact, I'm pretty sure no psychologist has ever prescribed these activities. But each of these three unusual instructions *is* directly inspired by practical recommendations taken straight from positive-psychology manuals. For example:

- Practice random acts of kindness twice a week. (The reward center of the brain experiences a stronger "dopamine hit" when we make someone else smile than when we smile first.)
- Think about death for five minutes every day. (Researchers suggest that we can induce a mellow, grateful physiological state known as "posttraumatic bliss" that helps us appreciate the present moment and savor our lives more.)
- Dance more. (Synchronizing physical behavior to music we like is one of the most reliable—not to mention the safest—ways to induce the form of extreme happiness known as euphoria.)

These three guidelines represent some of the most commonly prescribed happiness activities today; the first set of instructions just offers more *gameful* interpretations of them.

What, exactly, are happiness activities? They're like the daily multivitamins of positive psychology: they've been clinically tested and proven to boost our well-being in small doses, and they're designed to fit easily into our everyday lives. There are dozens of different happiness activities in the scientific literature to choose from in addition to the three listed above, ranging from expressing our gratitude to someone daily to making a list of "bright-side" benefits whenever we experience a negative life event. And they all have one thing in common: they are backed by multiple million-dollar-plus research studies, which have conclusively demonstrated that virtually anybody who adopts one as a regular habit *will* get happier.

Of course, if it were *that* easy, we'd all be a lot happier already. In fact, by nearly all measures, we're *not* substantially happier as a planet than we were before the rise of positive psychology in the 1990s. Rates of both clinical depression and mild depression globally are increasing so quickly, the World Health Organization recently named depression the single most serious chronic threat to global health, beating out heart disease, asthma, and diabetes.[1] In the United States, where we frequently put on happy faces for each other in public, we admit in private to surprisingly low levels of life satisfaction. In one recent nationwide survey, more than 50 percent of U.S. adults recently reported that they "lack great enthusiasm for life" and "don't feel actively and productively engaged with the world."[2] This is despite the fact that we have—more than ever before—better and wider access to evidence-based self-help tools in the form of best-selling positive-psychology books, not to mention countless magazine articles and blog posts.

So what's the problem? It turns out that *knowing* what makes us happy isn't enough. We have to act on that knowledge, and not just once, but often. And it's becoming increasingly obvious that it is just not that easy to put scientific findings into practice in our everyday lives.

We need help implementing new happiness habits—and we can't just help ourselves. In fact, when it comes to improving our collective happiness levels, self-help rarely works. Outside the structure and social support of a clinical trial or classroom, these self-help recommendations are surprisingly hard to implement on our own. Depending on the activity, we either *can't* or *won't* do them solo—and there are three big reasons why.

The first and most important reason is summed up best by Sonja Lyubomirsky, who laments, "Why do many of the most powerful happiness activities sound so . . . well, hokey?"[3] Lyubomirsky earned a million-dollar research grant from the National Institute of Mental Health to test a dozen different happiness activities—and she discovered that despite their incontrovertible effectiveness, many people resist even trying them. The most common complaint, according to Lyubomirsky? Happiness activities sound "corny," "sentimental," or Pollyannaish.[4]

"Such reactions are authentic, and I can't dispute them," Lyubomirsky admits.[5] We instinctively resist activities that feel forced and inauthentic, and many people are deeply suspicious of unadulterated feel-goodness. Shouldn't expressions of gratitude be spontaneous, not scheduled? Isn't it naive to constantly look for silver linings? What if I just plain don't *feel* like making a gesture of kindness today? When it comes to doing good and feeling good, we seem to think it's more "real" if we wait for inspiration to strike, rather than committing to doing it whether we "feel like it" or not. On top of that, there's just plain suspicion and skepticism of these unabashedly positive activities. There's an undeniable tendency toward irony, cynicism, and detachment in popular culture today, and throwing ourselves into happiness activities just doesn't fit that emotional climate.

Positive psychologist Martin Seligman explains that "the pervasive belief that happiness is inauthentic is a profound obstacle" to putting positive psychology into action.[6] Science just doesn't have a chance against our instinctive, visceral reactions—and, unfortunately, the best advice that positive psychologists have to offer seems to push all our cynical, skeptical buttons. For many people, happiness activities will need to be embedded in a more instinctively appealing— and less overtly do-good, feel-good—package.

The second obstacle to practicing simple happiness activities on our own is what I call the "self-help paradox." Self-help is typically a personal, private activity. When it comes to some activities—overcoming fears, identifying career goals, coping with chronic pain, starting a fitness routine—there's certainly reason to believe that self-help can work. But when it comes to everyday happiness, there's just no way personal, private activity can work—because, according to most scientific findings, there are almost no good ways to be happy alone for long.

As the author Eric Weiner, who has studied worldwide happiness trends, reports: "The self-help industrial complex hasn't helped. By telling us that happiness lives inside us, it's turned us inward just when we should be looking outward . . . to other people, to community and to the kind of human bonds that so clearly are the sources of our happiness."[7] Weiner makes an excellent point here: self-help isn't typically *social*, but so many happiness activities are meant to be. Moreover, positive psychology has shown that for any activity to feel truly *meaningful*, it needs to be attached to a much bigger project or community— and self-help just doesn't usually unfold collectively, particularly when self-help advice comes in the form of books.

Approaching happiness as a self-help process runs counter to virtually every positive-psychology finding ever published. Even if we can get ourselves past the hokiness problem, thinking of happiness as a self-help process will doom us to failure. Ideally, happiness needs to be approached as a *collective* process. Happiness activities need to be done with friends, family, neighbors, strangers, coworkers, and all the other people who make up the social fabric of our lives.

Finally, there's a self-help problem that isn't unique to the science of happiness: it's easier to change minds than to change behaviors. As Harvard professor of psychology Tal Ben-Shahar explains, we're often more willing to learn something new than we are to actively adapt our lives. "Making the transition from theory to practice is difficult: changing deeply rooted habits of thinking, transforming ourselves and our world, requires a great deal of effort," he writes. "People often abandon theories when they discover how difficult they are to put into practice."[8] Either we never try or we get bored or frustrated quickly.

Toward the end of his best-selling self-help book *Happier*, Ben-Shahar makes one last effort to convince people to make practical use of what they've read: "There is one easy step to *un*happiness—doing nothing." But unfortunately, that's exactly what most of us do after we read a book or magazine article about happiness: absolutely nothing. The written word is a powerful way to communicate knowledge—but it's not necessarily the most effective way to motivate people. We simply can't self-help our way out of the depression epidemic. Alongside platforms for communicating the science of happiness, we need platforms for *engaging* people in scientifically proven happiness activities.

And that's where ideas like sidewalk compliments, cemetery poker, and stationary dancing come in.

Shout compliments at strangers, play poker in a cemetery, and dance without moving your feet are all instructions from large-scale public games that I've designed specifically with the intention of creating opportunities for as many people as possible to participate in happiness activities they wouldn't ordinarily try.

These "crowd games"—meant to be played in very large, and usually face-to-face, groups—are called Cruel 2 B Kind, Tombstone Hold 'Em, and Top Secret Dance-Off. And they're all perfect examples of a new design practice called "happiness hacking."

The Invention of Happiness Hacking

The term "hacking" has its origins in the 1950s, when MIT students defined it as "creatively tinkering with technology."[9]

Back then, it was primarily radios that hackers were playing with, and it was a social activity: they would proudly show off their best hacks to anyone who would pay attention. Today, we most often think of hacking in the context of computing. You might associate the term "hacking" with malicious or illegal computer activity, but in the tech community it more commonly refers to clever, creative programming—especially if it takes a smart shortcut to accomplish something otherwise challenging. And as with the original

MIT hacks, there's still a tradition of showing off and freely sharing success-ful hacks.

Recently, especially in Silicon Valley circles, "hacking" has been used more broadly to talk about a kind of creative, hands-on problem solving that usually, but not always, involves computers. A good example of this phenom-enon is a movement called "life hacking." Life hackers look for simple tips and tricks to improve productivity in everyday life—such as adopting the "ten/ two rule." The ten/two rule means you work for ten minutes, and then let yourself do something fun and unproductive for two minutes—checking e-mail, getting a snack, browsing headlines. The theory is that it's easy to com-mit your attention to work for just ten minutes at a time, and as a result you'll get fifty good working minutes out of every hour. For many people, that's a huge boost in productivity. To make it easy to adopt this habit, life hackers have created desktop and mobile phone applications that buzz alternately every ten and two minutes to keep you on track.

Life hacking positions itself in direct contrast to self-help; it's meant to be a more collective way of working out solutions and testing them out together. As Merlin Mann, one of the leading life hackers, explains, "Self-help books tend to be about lofty ideas, whereas life hacks are about getting things done and solving life's problems with modest solutions."[10] Any good hack—whether it's a computer hack or a life hack—should be free to adopt and extremely *lightweight*—meaning easy and inexpensive to implement—without any spe-cial equipment or expertise.

It was in this spirit that I coined the term "happiness hacking" several years ago.[11]

Happiness hacking is the experimental design practice of translating *positive-psychology research* findings into *game mechanics*. It's a way to make happiness activities feel significantly less hokey, and to put them in a big-ger social context. Game mechanics also allow you to escalate the difficulty of happiness activities and inject them with novelty, so they stay challenging and fresh.

When I design games today, I always embed at least one proven happiness activity into the game mechanics—and sometimes I invent games based entirely on a handful of new research findings. It's my way of enacting the tenth fix for reality:

 FIX #10: HAPPINESS HACKS

Compared with games, reality is hard to swallow. Games make it easier to take good advice and try out happier habits.

HAPPINESS HACK #1: UNLOCKING
THE KINDNESS OF STRANGERS

The two most frequently recommended happiness activities across the scientific literature are to express gratitude and practice acts of kindness. Recent research has shown that we don't even have to know someone to experience the benefits of thanking and being nice to them. Even fleeting acts of gratitude and kindness toward strangers can have a profound impact on our happiness. And positive gestures from strangers can make a big difference in how rich and satisfying our everyday lives feel.

Sociologists call the positive relationships we have with strangers "transitory public sociality." We experience it in all kinds of public places: sidewalks, parks, trains, restaurants, stadiums, and coffee shops, for example. These transitory social interactions, when they happen, are usually brief and anonymous: we catch another's eye, we smile, we make room for someone else, we pick up something someone has dropped, we go on our way. But these brief encounters, taken cumulatively, have an aggregate impact on our mood over time.

Researchers have shown that sharing the same space for even just a few minutes a day with kind and friendly strangers makes us more optimistic, improves our self-esteem, makes us feel safer and more connected to our environment, and generally helps us enjoy our lives more.[12] And if we return the favor, we benefit as well: when we give to others, or act cooperatively, the reward centers of the brain light up.[13]

But strangers aren't always inclined to be friendly to each other—and some researchers believe our shared spaces are becoming less friendly over time.

Dacher Keltner has devised a simple way to test this theory: a mathematical method for measuring the social well-being of any shared environment. It's called the "*jen* ratio," from the ancient Chinese word for human kindness. It compares the total positive interactions between strangers to the total negative interactions, in a given period of time and in a given place.[14] The higher the ratio, the better the social well-being of the space and the happier you're likely to feel after spending time in it. The lower the ratio, the poorer the social well-being, and the unhappier you'll be if you spend too much time there.

To measure the jen ratio of a space, you simply watch it very closely for a fixed period of time—say, one hour. You count up all the positive and negative microinteractions between strangers, keeping track of two different totals: how many times people smile or act kindly toward each other, and how many times people act unfriendly, rude, or openly uninterested. All the positive microinteractions— such as big smiles, a hearty thank-you, a door being held open, a concerned question—get tallied on the left side of the ratio. All the negative microinteractions—a sarcastic comment, an eye roll, an unexcused bump, someone cursing under their breath—get tallied on the right side.

The jen ratio is a simple but powerful way to predict whether being in a particular place will make us happier or unhappier. When Keltner surveyed several years' worth of recent research on social well-being and social spaces, he concluded, "Signs of a loss of *jen* in the United States are incontrovertible . . . with a jen ratio trending toward zero."[15]

So how can we raise the jen ratio of everyday shared spaces? The solution is obvious, if hard to enact: we need to convince large numbers of people to do things like smile more, be more welcoming, express more gratitude, or pay more compliments. Positive psychologists, of course, have already given us this recommendation—but, as Lyubomirsky's research shows, such recommendations rarely inspire direct individual action. Who wouldn't feel daunted by the challenge of trying to increase the jen ratio of a big public space single-handedly? More likely, it would take a crowd, and not a single person, to effectively bump up the jen ratio. But there simply aren't any well-established social traditions for going out and expressing gratitude or being kind to strangers together.

As a game designer, it is clear to me that we can tackle these problems by making this behavior more challenging and social. All it needs are a few arbitrary limitations, some multiplayer obstacles, and a feedback system in order to turn unlocking the kindness of strangers into a game.

So what exactly would a kindness game look like? And who would play it? These are questions I asked myself a few years ago, and, together with my good friend and fellow game developer Persuasive Games cofounder Ian Bogost, I decided to invent a game with the core mechanism of performing acts of kindness on strangers—as *sneakily* and *stealthily* as possible.

It would work just like the popular college campus game Assassins, in which players are assigned targets via e-mail, and then proceed to stalk each other across campus for days or even weeks to eliminate their targets with water guns and other toy weapons. But in our version, the game would be shorter (an hour or two) and confined to a much smaller space (a few city blocks, a park, or a large public plaza). And players wouldn't kill each other with toy weapons—they'd kill each other with kindness. Most importantly, they wouldn't be given specific targets, so anyone nearby was fair game for a thank-you or a compliment. And instead of being eliminated from the game when "killed," players would join forces and cooperate with each other to keep performing bigger and more spectacular acts of kindness.

We called it Cruel 2 B Kind, or C2BK for short, after the famous line from Shakespeare's tragedy *Hamlet*. We debuted it in 2006 in San Francisco and New York City; it's since been played everywhere from Detroit, Michigan, and Johannesburg, South Africa, to Stockholm, Sweden, and Sydney, Australia. Here's how it works:

Cruel 2 B Kind is a game of benevolent assassination. At the beginning of the game, you are assigned three secret weapons via e-mail or text message. To onlookers, these weapons will appear like random acts of kindness. But to other players, the friendly gestures are deadly maneuvers that will bring them to their knees.

Some players will be killed by a compliment. Others will be slain by a smile. You and your partner might be taken down by a happy offer to help.

You can attempt to kill anyone else who is playing the game. However, you will have no idea who else is playing the game. You will be given no information about your targets. No names, no photos—nothing but the guarantee that they will remain within the game boundaries during the designated playing time. *Anyone* you encounter could be your target. The only way to find out is to attack them with your secret weapon.

Watch out: the hunter is also the hunted. Other players have been assigned the same secret weapons, and they're coming to get you. Anything out of the ordinary you do to assassinate *your* targets may reveal your own secret identity to the other players who want you dead. So be cool when you attack. You don't want to alarm innocent bystanders . . . or give away your secret identity.

In many cases, you and another player will spot and attempt to kill each other at the same time! For this reason, the weapons are assigned powers according to the classic rock-paper-scissors model: a hearty welcome beats a thank-you, for example, or a killer compliment beats a wink and a smile. And if both players deploy the same weapon at the same time? It's a standoff—you turn and run in the opposite direction, and both players must wait thirty seconds before attacking again. As targets are successfully assassinated, the dead players join forces with their killers to continue stalking the surviving players. The teams grow bigger and bigger until two final mobs of benevolent assassins descend upon each other for a spectacular, climactic kill.

Will innocents be caught in the cross fire? Oh, yes. But when your secret weapon is a random act of kindness, it's only cruel to be kind to other players . . .

A team of C2BK players in London.
(Alex Simmons for the Hide & Seek Festival, 2008)

In addition to this basic rule set, we created a database of possible weapons, and invited players to suggest their own. For example:

- Welcome your targets to beautiful [your neighborhood or city].
- Tell your targets, "You look gorgeous today!"
- Point out something amazing to your targets, such as, "Isn't that an amazing bird!"
- Praise your targets' shoes.
- Offer to help your targets with something specific.

- Thank your targets for something they're doing right now.
- Express "mind-boggling" admiration of your targets.
- Wink and smile at your targets.
- Volunteer to answer any questions your targets have about something specific nearby.

Besides swapping kind gestures for toy weapons, the two most important design decisions that we made were to shrink the window of play and to obscure the number and identity of players. In a regular game of Assassins, the game is too spread out physically and time-wise to have a significant impact on the local environment. By reducing the field and length of play, we "concentrated" the game to increase its impact and intensity. And in a traditional game of Assassins, players know exactly who they're targeting. Bystanders *do* occasionally get caught up in the cross fire, but it's always an accident, and it's usually not fun for the victim. (No one wants to be unexpectedly splattered by a water gun if they're not participating in a game!) In C2BK, however, we wanted bystanders to get hit—every positive microinteraction would increase the jen ratio, regardless of whether it improved the player's score. In fact, the higher percentage of "misfires" (i.e. toward nonplayers), the better.

To be fair, being accidentally "attacked" by a player is somewhat startling— but also potentially enjoyable. In a best-case scenario, the "victims" of play feel genuinely welcome or complimented or appreciated. At the start of the game, when players are timid and groups are small, this tends to be the case. Later, as the players get bolder and teams get larger, strangers are more likely to be clued in to the unusual nature of the activity and provoked to wonder why everyone is making such showy efforts of gratitude and kindness. This is one of the intended effects of the game—to reveal if friendly gestures are considered out of place, and to provoke people to wonder why exactly that is. Of course, by the end of a game, being complimented by a horde of twenty or more adrenaline-pumping players is clearly no everyday act of transitory public sociality. No one is likely to mistake *that* for an ordinary act of kindness. But the spectacle works toward a different positive end: it adds a spark of novelty and curiosity to the environment. It's bracing, but benevolent—and

our goal in including this level of spectacle was to jolt people out of their social bubbles.

Years of low jen ratios may make some bystanders more cynical and jaded
than others—and for them, getting welcomed, serenaded, thanked, or complimented by a single stranger or a crowd of strangers might not initially be a positive experience. That's why we were careful to playtest the various "weapons," to
whittle the list down to the most consistently positive-reaction-provoking gestures. I've also observed—and filmed—many C2BK games in action, specifically
looking for signs that the majority of bystanders benefit, in addition to the players.
To date, my studies have shown that the visible positive reactions—smiles, wide-
eyed curiosity and amazement, cheerful replies—far outnumber the blank stares
or negative reactions.

Ultimately, though, it's the players who benefit most from the game. That's
because when you play C2BK, the basic happiness activities of expressing
gratitude and practicing random acts of kindness are made more engaging.

First of all, the C2BK game makes the kindness activities more interesting.
There are two obstacles in the way of your performing them: you don't know
who to attack, and you're trying to sneak past and avoid other players. Much
of the game is spent scouring the environment for targets while trying to keep
a low profile. You can't help but wonder about everyone you see: Are they
playing the game, too? Strangers become potential targets and allies, and the
only way to find out if they share your secret is to interact positively with them.

C2BK also produces adrenaline. Paying a compliment becomes an act of
courage: you have to work up your nerve to overcome the social norms of
ignoring strangers, and you have to do it as quickly as possible, because every
second that passes is a second that another player could be targeting you.
C2BK also has more pronounced fiero moments. Players and teams let out
big hollers and cheers when they've made a successful kill, and the fiero moment is intensified by the number of misfires you've made on the way. My
rough estimate from observing several games is that C2BK participants attack
on average five times as many nonplayers as players.

The game also has more novelty than ordinary acts of kindness. It encourages you to think about being nice to strangers in different environments—

and the possibilities are endless. It's most frequently played in downtown settings, but Cruel 2 B Kind isn't just a game for sidewalks and parks—any public or shared space could benefit from having its jen ratio raised. I've received reports of C2BK games played in settings as diverse as high-rise office buildings, arts festivals, libraries, shopping malls, convention centers, apartment complexes, college dorms, public train systems, and even the beach.

Finally, C2BK gives you collaborators in your happiness activity. You can gather up your friends to be on a team with you, and as you start getting folded into larger and larger groups—the biggest C2BK game I've participated in had more than two hundred players in a three-by-three city-block radius—you build up a sense of being on a collective mission to kill with kindness. It's the kind of emotionally charged experience that can forever change how you see your own kindness capabilities. Even if you play C2BK formally only once or twice, you may find yourself continuing to think of friendly gestures as secret weapons you can deploy anytime, anywhere. (This is exactly what players report to me weeks and months after their first time playing the game.) The game gives you a different view into two happiness activities, charging them with more excitement, fiero, and social energy.

CRUEL 2 B KIND, like many happiness hacks, isn't a product. There's no software to download, no license to buy, no fee to pay. It's meant to be a solution to a problem—the problem of how to increase the jen ratio of a shared space—and it can be adopted and adapted by anyone, anywhere. It was cheap to invent—Ian Bogost and I worked for free, and the whole project probably cost us less than five hundred dollars in expenses to playtest and launch.

The game can be played using any kind of mobile communications technology: text messaging, mobile e-mail, and Twitter are the most popular platforms for C2BK.

To help spread the hack, the Cruel 2 B Kind website includes a few essential tools. There's a six-minute video showing the highlights of a game from start to finish, to help potential players get up to speed quickly. There's also a one-page "cheat sheet" with rules and frequently asked questions that players can print out and bring to the game.

It's hard to keep track of all the C2BK games that happen—game organizers don't have to get our permission to run a game, so I rely on voluntary reports. Three years after launching the game, I still hear from new game organizers roughly every month. At the very least, C2BK has been played in more than fifty different cities, in ten countries, on four continents.

Recently, I received news of perhaps the most interesting C2BK setting yet: Summer Darkness, one of the biggest gothic festivals in Europe. The three game organizers wrote me an e-mail from the festival's home city, Utrecht, in the Netherlands, explaining, "Ultimate goal: get the Goths (coming from all over Europe), and 'civilians' (non-Goths) to play together in the streets."

Now, if any group would find straightforward happiness activities hokey, I'm pretty sure it would be goths. The gothic subculture, of course, is known for embracing dark, mysterious, and morbid imagery. There's a kind of loneliness and alienation deeply entrenched in gothic stories, music, and style. And Summer Darkness is officially billed as a "dark underground lifestyle" festival, so it might be the last place you'd expect to see people throwing themselves into extroverted interaction with strangers, let alone cheerful expressions of gratitude and random acts of kindness.

For Cruel 2 B Kind to be an appealing activity to this community stands as excellent proof, I think, that even the most unabashedly do-good activity can be transformed into mischievous fun. It's proof that happiness hacking works. You really can turn positive-psychology advice about what's "good for you" into something that you really *want* to do.

HAPPINESS HACK #2:
PLAYING OUR RESPECTS

Tombstone Hold 'Em is a variation of Texas Hold 'Em poker designed to be played in cemeteries. It is also, without a doubt, the most controversial game I've ever designed.

To say that some people find the idea of playing games in a real-world cemetery inappropriate would be putting it mildly. In the United States in particular, we have a culture of grieving as quietly, as privately, and as solemnly as possible. Cemeteries—despite having been popular as public parks and recreation spaces in the nineteenth and early twentieth centuries—today are largely single-use memorial spaces. They're meant to be briefly occupied by mourners first and foremost. Some older or more scenic cemeteries may draw other visitors, but they generally move through them as inconspicuously as possible.

But I've never been prouder of a game design, and for one reason: players widely report being able to think about death and lost loved ones in a more positive way after playing Tombstone Hold 'Em. And that's the point of the game. It's a happiness hack meant to create more social, and more enjoyable, ways of remembering death.

Thinking about death is one of the most highly recommended happiness activities, but it's also one of the most difficult to convince ourselves to undertake. We're accustomed to pushing thoughts of death out of our minds, not cultivating them. Tombstone Hold 'Em is meant to make remembering death easier and more rewarding, by taking advantage of the largely underutilized social and recreational potential of cemeteries.

The central activity of Tombstone Hold 'Em poker is learning how to "see" a playing card in any tombstone, based on its shape (the suit) and the names and date of death (the face value). Once you can read stones as cards, you can spot "hands" all around you. The game works in any cemetery, as long as there are clearly marked tombstones. Here's how it plays out:

The key to understanding Tombstone Hold 'Em is that there are only four shapes you get on top of a tombstone. Pointy equals spades, statue on top equals clubs, rounded equals hearts, flat equals diamonds. That's how you tell the suit.

Now take the last digit in the year of death. That's your face value. Died 1905—that's a five. Died 1931—the one is the ace.

But two names on the stone? Ignore the dates—that's a jack. Three names is a queen. Four or more names is kings.

Now maybe you have to clear away some leaves or dirt or litter in order to read the cards. That's good—it helps keep those old stones taken care of. Just be gentle with 'em.

Now, for a hand. You play it like regular Texas Hold 'Em, but in reverse. First, lay out the whole "flop" upfront. Five regular playing cards. Now everyone antes up, and then each pair (you've got to have a partner to play) has three minutes to find their two best hold cards.

You can pick any two cards you want from anywhere in the cemetery—but you have to use the stones, not regular playing cards this time. The trick is you have to be able to *touch* both tombstones and each other at the same time. So maybe I've got a hand on a ten of hearts and the other on my buddy's toe, while he's stretched out to touch another heart for the flush. If we can't make the reach, we can't claim the cards.

So find any pair you like and put a pair of poker chips on 'em to claim 'em. Now no other pairs can pick your two stones for their own pocket pair.

Two players show their best hand during a Tombstone Hold 'Em game in the historic Congressional Cemetery in Washington, D.C.
(Kiyash Monsef, 2005)

All this has to happen fast, because after three minutes whoever's got the working watch yells out, "Last call!" and everyone runs hell for leather back to the flop and says what they found. Only the best hand has to prove it, and winner takes all the antes. In the case of ties, first back to the flop is the winner.

One more thing: no betting or bluffing in traditional Tombstone Hold 'Em. Only way to win is to earn it. So go out there with your partner and make sure you find the best pair.[16]

Tombstone Hold 'Em allows players to actually get to know the people at rest in the cemetery. You read the stones, you learn the names, and you start to wonder about their stories—because every time you pick a pocket pair, you're recruiting two dearly deceased as allies in the hand. Playing the game in a perfectly manicured cemetery is good, but playing it in a cemetery that could use a little loving care is better—it's more challenging, and more rewarding. As you clear away clutter from the stones to make them legible again, you're not just playing in the cemetery—you're taking care of it.

The game is meant to be played by at least four people, and ideally in larger groups—the larger the group, the more enlivened the cemetery feels. I've organized large-scale Tombstone Hold 'Em games in historic cemeteries in Kansas City, Atlanta, New York City, Los Angeles, and San Francisco, for crowds ranging between twenty and two hundred. With a group that big, you've got a dozen or more "flops" going at the same time, spread out around the cemetery, on various benches, tree stumps, or mausoleum steps. Whenever I've organized a larger game, I've done so with the official permission and assistance of the cemeteries. But I've also played much smaller, unofficial games everywhere from Austin to Helsinki to Barcelona to Vancouver. If it's a small group—say, four, six, or even eight players—it doesn't raise too many eyebrows, especially if you're sure to play well out of the way of anyone who might be visiting the cemetery for more traditional purposes.

But before I get too much further dissecting the experience of playing Tombstone Hold 'Em, perhaps I should explain how I came to be designing a crowd game for cemeteries—and how happiness research convinced me to take on such an unusual project.

In 2005, I was working as a lead designer for a game company called 42 Entertainment. We accepted a commission to develop an alternate reality campaign for the Western-themed video game *Gun*, developed by Neversoft and published by Activision. The goal of the alternate reality campaign was to give gamers the chance to directly experience the historical world of *Gun*, the American Old West of the 1880s. The centerpiece of the alternate reality campaign was an online poker platform, styled in a Western theme. Gamers were invited to compete in online Texas Hold 'Em tournaments set in the past, competing at the same table with historical characters from the 1880s. It was a unique combination of historical role playing and card playing.

Alternate reality games usually have a real-world component, and since *Gun* featured real-world characters who'd died in the Old West, we came up with the idea of using real cemeteries as a site for some kind of live-action experience. Because of my expertise in running reality-based games, I was put in charge of figuring out what the live cemetery events would be.

On one hand, I was excited by the concept. In a world where video gamers are much maligned for being desensitized to violence, it struck me as a particularly provocative idea to send gamers to the *real-world graves* of characters they had killed in *Gun*. But I also felt some trepidation, hitting up against the cultural norms involving cemeteries. I *really* didn't want to organize some kind of rowdy, unauthorized "flash mob," so I started researching historic cemeteries and brainstorming what kinds of things gamers could do in them.

One of the first things I discovered was that cemeteries in the United States were absolutely *desperate* to convince people to spend more time in them. According to cemetery industry statistics, the average grave receives just *two* visits in its lifetime—*total*, by *any* friend or family member—after the initial flurry of visits that immediately follows the burial.[17] We think of cemeteries as spaces for mourning, but the truth is, mourners do not regularly return. Meanwhile, others are generally discouraged by social norms from spending significant time in the space—it's considered either unseemly or morbid.

As a result, cemeteries are for the most part empty. And lack of participation in cemetery spaces has become a huge problem from an industry standpoint (cemeteries are running out of money), a community standpoint (the less visited a cemetery is, the more likely it is to be poorly maintained and vandalized), and, perhaps most of all, from a happiness standpoint (according to research, the less time we spend in cemeteries, the more likely we are to suffer from fear and anxiety about death).

I was first tipped off to these problems by a *New Yorker* article about the decline of American cemeteries, which was published while I was in the midst of my research. In the article, Tad Friend documents how Americans today spend less time in cemeteries than ever before, despite the vast expanses of green space they take up and the escalating costs of maintaining them. "Who are cemeteries for? The living or the dead?" he asks. We've apparently convinced ourselves they're for the dead, since we don't visit them. But that's ridiculous, as Tad Friends argues: "They're for the living; the dead can't enjoy them. The trick for cemeterians is to get the living to come to them." He documents a range of fledgling efforts on the part of cemeteries nationwide to become more relevant to the living. There are, for example, movies projected at night on the sides of mausoleums in Hollywood, 5K graveyard races in Kansas City, and dog-walking clubs in historic Washington, D.C., cemeteries.[18]

As I researched the subject further, I discovered that many cemeteries were fighting for their very survival, and largely as a result of the American desire to keep the reality of death as far removed from our daily lives as possible. For decades private cemeteries have been quietly sold to accommodate new highways, schools, and condominiums; the graves are typically relocated to more remote areas. Meanwhile, many public and historic cemeteries receive insufficient funds to maintain the grounds properly; with such low visitation rates, they have a hard time documenting their value to the community. And abandoned cemeteries once belonging to now defunct churches are being adopted by local community groups in an effort to repair them and preserve their historical value.[19]

People who care about and run cemeteries make many good arguments in favor of protecting them: they are a unique repository of historical data, they have significant architectural value, and, not least of all, there is the ethical imperative to honor contracts to families who have buried their loved ones with the expectation that they will be cared for in perpetuity.

All of these are worthy reasons to enliven cemetery spaces today—but what really convinced me was the happiness research.

In his report on global happiness trends, Eric Weiner writes that death is "a subject that, oddly, comes up an awful lot in my search for happiness. Maybe we can't really be happy without first coming to terms with our mortality."[20] It's a strange idea, but it's not a new one. In *The Happiness Myth*, happiness historian Jennifer Michael Hecht devotes an entire chapter to "the age-old advice to remember death, to keep it in the forefront of our minds for the sake of bettering the life we lead now."[21] She traces the idea all the way back to Plato, who advised students to "practice regular meditation upon death," and to Buddha, who said, "Of all mindfulness meditations, that on death is supreme." Even Epicurus, the ancient Greek philosopher best known for encouraging followers to seek simple pleasures, put death at the center of his vision of happiness, arguing that it is only when we shake free our fear of death that we can truly enjoy life.

According to Hecht, since ancient times meditations on death have served the same purpose: to replace fear and anxiety with a kind of calm, mellow gratitude for the life that we're given. And today, these traditions have the backing of contemporary science. Positive psychologists have found that grappling with the reality of death forces a kind of mental shift that helps us savor the present and focus our attention on the intrinsic goals that matter most to us. Hecht has coined a term for this realignment of priority and attention: posttraumatic bliss. "There are feelings in this life—good and bad—that cannot be conquered by intellect or force of will," she writes. "Almost dying can realign you in a way that is the positive incarnation of trauma: posttraumatic bliss."[22]

Researchers have documented the phenomenon of posttraumatic bliss among patients confronting a terminal medical diagnosis. Something seems to click in their minds, empowering them to enjoy their lives more. It's not just that they've realized how precious life is; there seems to be some kind of significant mental clearing that occurs along with a new ability to focus on positive goals. In *Happier*, Tal Ben-Shahar quotes Irvin D. Yalom, a psycho-therapist who often works with dying patients: "They are able to trivialize the trivial, to assume a sense of control, to stop doing things they do not wish to do, to communicate more openly with families and close friends, and to live entirely in the present rather than in the future or the past."[23] This rare and intense positive focus on getting the most out of life is hard to come by in our ordinary lives, Ben-Shahar notes—especially when we spend so much time collectively trying to avoid thinking about death.

Can we learn to savor life and achieve that intensity of positive focus with-out the trauma of a terminal diagnosis or a near-death experience? That seems to be the idea behind classical and religious advice, and today positive psy-chologists like Ben-Shahar recommend activities such as imagining ourselves on our deathbeds as a way to try to provoke this positive clarity.

But as a happiness activity, solitary deathbed reflection leaves a lot to be desired. It's simply not something most of us are inclined to do—or if we are, we're not likely to take it seriously or do it for very long. We can't just tell ourselves to remember death—the ancient philosophers, Hecht notes, insisted that "it takes active meditations and gestures."[24]

Moreover, it's hard to force ourselves to grasp the reality of our own indi-vidual mortality. It's easier to acknowledge the universality of death—and that's where cemeteries come in. Cemeteries present us with vivid, extreme-scale, irrefutable historical evidence of the one thing that connects us all, the one thing that makes it possible to enjoy life to the fullest—if only we felt inclined to spend more time in them.

At this point in my research, I was convinced that spending more time in cemeteries was a worthy social goal—and that a graveyard game could do a lot more than bring a historical video game to life. The *Gun* project was the per-fect opportunity for a happiness hack. And the key to making this happiness hack work would be to generate the kind of positive emotions we typically as-sociate with crowd games—excitement, interest, curiosity, social connection—and simply unleash it in the physical context of a cemetery.

Once I started playtesting in cemeteries, the design pieces fell quickly into place. I knew I would need a focused activity that, in some respect, had nothing to do with remembering death — that had to be the side effect, not the purpose, of the game. And since Texas Hold 'Em poker was a larger theme for the alternate reality campaign, it made perfect sense to bring the familiar game to the cemeteries.

But the poker needed to be site-specific and really needed to use the natural affordances of a cemetery — otherwise, you'd just play the game somewhere else, defeating the entire purpose. And that's where the idea of using stones as playing cards came in. Tombstones are the single design feature that all cemeteries have in common, guaranteeing the game would be playable anywhere. And paying close attention to the content of the tombstones directly supported the goal of the happiness hack — each card you "decoded" meant literally staring death in the face, but in a way that wouldn't provoke fear or anxiety.

As for the other design choices, I made it a partner game because this seemed like a good way to ensure that it was not just social, but also cooperative. Cooperation always provokes positive emotion and meaning in games, especially if a physical connection is involved. Meanwhile, touch is one of the fastest ways to build social bonds — holding hands, touching someone's back, and patting a shoulder all release the oxytocin chemical that makes us like and trust each other. But, as Dacher Keltner's positive-emotion research has shown, "We live in a touch-deprived culture."[25] To put it another way, as Michelangelo said, "to touch is to give life" — and I couldn't think of a better way to enliven a cemetery than to unleash a flood of oxytocin in the crowd.[26]

When a game is in motion, there's an air of happy participation that simply isn't the norm for cemeteries. It's a distinct break from the typical atmosphere, usually one of quiet, solitary reflection or collective mourning. At the same time, small pockets of conversation often break out, among friends as well as strangers — people sharing small pieces of their own experiences with mourning and loss. This has unfolded at every Tombstone Hold 'Em game I've been to — it's almost impossible not to, given the setting. In this way, the game perfectly serves its purpose: it simultaneously activates positive emotions and social bonds while putting us in the perfect environment to get our recommended daily reminder that we are all dust, and to dust we shall return.

Which brings us back to the potential controversy. Tombstone Hold 'Em was featured in a number of news articles, and some online readers commented that the game seemed "disrespectful," "insensitive," or even "obscene." Which raises the question: Is it appropriate to play games in a cemetery? Based on my direct experiences, absolutely. At more than a dozen trials I've organized of Tombstone Hold 'Em, participants have overwhelmingly agreed that this particular game *feels right* in the space—especially when the net result is that the tombstones receive more attention from the living and are better taken care of as a result.

Perhaps more than any other project I've worked on, Tombstone Hold 'Em has demonstrated one of the most vital powers of gameplay: it gives us explicit permission to do things differently. We are accustomed to being asked to behave and think unconventionally in a game. We're used to being creative and playing outside of social norms when we're inside the socially safe "magic circle" of a game. And the more people who come together to play an unconventional game like Tombstone Hold 'Em, the safer it feels. A crowd carries the social authority to redefine norms.

Does it really work as a happiness hack? I've played Tombstone Hold 'Em with hundreds of people and spoken with nearly every one of them about it immediately afterward. (The games are usually followed by social gatherings in restaurants or bars, a way to decompress after what can be an intense, emotional experience.) The most common reaction is that players felt "more comfortable" being in the cemetery after playing. Other words most commonly used to describe the experience were "strangely happy" and "relaxed," as well as "grateful" and "connected" to the people at rest. I've even talked to visitors at the cemeteries who spotted some of our players in the distance and asked me about the game; just once did a visitor express dismay. Most often, I heard a variation of the following sentiment: that it's nice to see a loved one's final resting place not lonely and empty, but full of people running, smiling, laughing, and having fun together.

Since I first shared the rules online, the game has spread mostly by word of mouth—like most good hacks—and I periodically hear about Tombstone Hold 'Em games popping up in cemeteries around the world. It's the best outcome possible for any happiness hack: a solution that's been tested, proven, and shared, and now continues to be passed around to those who can benefit from it. Today, Tombstone Hold 'Em lives on as a viral happiness solution— it's free to adopt or adapt, and no products or special supplies or technology is required. All you need are a set of regular playing cards, something to use as poker chips (some people use coins or colored glass stones), and a way to invite friends or strangers to play their respects with you.

HAPPINESS HACK #3:
ACTIVATING THE DANCE SECRET

"As a happiness lesson, nothing could be more straightforward: if you get a chance to dance in a circle, get up out of your chair and do it."[27] That's Jennifer Michael Hecht's advice in *The Happiness Myth*, and with good reason. Dancing together has been used throughout human history as a reliable source of a special kind of euphoria, the *dancer's high*.

Dancer's high is what we feel when endorphins (from the physical movement) combine with oxytocin (from touch and synchronized movement) and the intense stimulation of our vagus nerve (what we feel when we "lose ourselves" in the rhythms of the music and are part of a crowd moving together). It's an expansive mixture of excitement, flow, and affection that is hard to experience any other way.[28]

But dancing in groups also makes many people feel embarrassed or awkward. Everything from self-consciousness to social anxiety to a general disdain for any kind of group participation can prevent us from joining or fully enjoying a dance.

To really dance your heart out in front of others, to not hold back at all, is a daunting proposition for many (although certainly not all) people. It requires letting go, and showing people a side of yourself—exuberant, unguarded— that you might ordinarily keep hidden. For some, revealing that side requires a lot of trust in the people around you. And, in fact, according to positive-psychology researchers, the necessity of trust is one of the reasons why dancing is such a powerful happiness activity.

When we dance, we're forced into an emotionally and socially vulnerable state in which we have to hope and trust that others will embrace us, rather than judge us. At the same time, we're given the opportunity to embrace others and help them feel more comfortable dancing. In other words, dancing with others is a chance both to receive and to express our compassion, generosity, and humanity. As a result, Dacher Keltner writes, "Dance is the most reliable and quickest route to a mysterious feeling that has gone by many names over the generations: sympathy, agape, ecstasy, jen; here I'll call it trust. To dance is to trust."[29]

But first, we have to have both the *desire* to dance and the *nerve* to do it. Many of us are missing one or the other.

Some people, as a rule, just don't like "getting involved." And group dances in particular set off all kinds of hokiness alarms. It's no coincidence that one of the best-known group dances is actually called the "hokey pokey." If you're not in the mood for dancing, when a big group dance breaks out at a wedding or a street festival, for instance, being dragged into it can feel incredibly forced and inauthentic.

Others have the desire, but simply lose the nerve.

The more I kept hitting up against the same happiness advice—dance more, in large groups if possible—the more convinced I became that there had to be a way to make it easier for introverted types who were more likely to watch from the sidelines to participate, and to give people who are already willing to dance together more daily opportunities. After all, even people who are ready and willing to dance anytime, anyplace, don't get nearly enough opportunities. We simply don't have a lot of everyday venues for dancing together. I began to wonder: how could we all sneak a few minutes dancing together into our *everyday* lives, not just the occasional weekend?

My solution: take all the basic mechanics of a massively multiplayer online role-playing game, and swap in real-life *dance quests* and *dance-offs* for traditional role-playing quests and raids. I called it Top Secret Dance-Off, or TSDO for short, and I launched it as a stand-alone social network site dedicated to the adventure of dancing together.

ADVENTURERS WANTED.
NO DANCING TALENT REQUIRED.

Welcome to Top Secret Dance-Off, an underground network of otherwise ordinary people seeking to activate the dance secret—an elusive power said to be hardwired into our brains, and requiring highly unusual dance experiences to unlock it.

Top Secret Dance-Off is an adventure you can undertake anywhere in the world. No dance skills or talent is required. In fact, you may find yourself rewarded more for awkward dancing than for a virtuoso performance. Activating the dance secret isn't about being a good dancer. It's about being a clever dancer, a brave dancer, and occasionally a stealthy dancer.

Adventures will involve undertaking a variety of challenging, top secret dance missions on video, sometimes in the privacy of your home, and sometimes in the most unlikely environments. You may play alone, or with your friends. Mask wearing or other disguises are required.

As you try to activate the elusive dance secret, you'll earn points by completing dance quests and participating in dance-offs. As you earn more points, you'll level up. The higher the level, the more dance secret you've activated.

For every quest you complete and every dance-off you enter, you'll also be earning choreopowers, such as style, courage, humor, and coordination. Your choreopowers reveal *your* personal strengths as a top secret dancer—and all choreopowers are awarded by other members of TSDO, in the comments on your videos.

Full activation of the dance secret occurs at level 100. How do you get there? Just complete twenty-one increasingly challenging quests and win at least a dozen dance-offs, and you'll be fully activated for life.

Dance Quest #1: Disguise yourself.

Your first mission is to disguise yourself. After all, this *is* top secret.

But we're not talking a full disguise. That would make dancing . . . difficult. So, to keep your TSDO identity a secret from the rest of the world, you must create a lightweight disguise that *covers at least part of your face*. It might be a mask, a scarf, modified sunglasses, face paint, a wig, or . . . ? It's your face. You decide how to hide it. But make sure you really like your disguise—because you'll need to don the exact same one for *all* future quests and dance-offs. Batman and Wonder Woman didn't make new suits every time they went out to save the world, did they? So pick something you like, and stash it somewhere safe—and secret. You'll be needing it.

Now: *Make a video* introducing yourself to the TSDO world. You must be 1) wearing your disguise and 2) dancing. Pick any song you like. BUT—and here's the tricky party—*keep your secret weapons in check* for now. That means *no moving your feet*. Dance, but don't move your feet. Like they're locked in cement. Got it?

Keep the video short—less than thirty seconds. Upload it to the TSDO site when you're ready to unleash your top secret dance identity and start earning your choreopowers.

Admittedly, this is not exactly dancing together, at least not in the traditional sense. Most of the dance quests and dance-offs involve dancing alone, then uploading a video to the Top Secret Dance-Off social network site. But the game serves two important purposes toward making it easier to dance together.

First, by providing a goal-oriented, feedback-rich, obstacle-intensive environment for dancing, it makes dancing more motivating, fun, and addictive. In other words, it increases a person's likelihood of dancing at all. Second, TSDO puts dancing, even dancing around your apartment alone, into a collective social context. It still takes nerve to share your dancing with an online community—and it's a real opportunity to express compassion, generosity, and humanity when we cheer on other players in the comments. In other words, the game is a hack for group dancing—a way to dance together alone, and make people more likely to dance together for real, in the future.

The heart of the TSDO experience is the never-ending list of potential dance quests, each of which adds a unique, unnecessary obstacle to dancing. By putting an obstacle in your way, TSDO makes it much harder to be self-conscious about dancing: you're focused on completing the challenge, not necessarily on how you look. It also gives you permission to dance badly, by restricting "normal" ways of dancing. The first dance quest—to dance without moving your feet—is a perfect example of this design strategy in action: it automatically rules out pretty much any kind of traditional or obvious dancing. Excelling at stationary dancing requires silliness, creativity, or just plain enthusiasm—not necessarily grace, sexiness, strength, or whatever else we might associate with natural dance talent.

Other dance quests include missions like "Dance upside down," "Dance in a crosswalk," "Dance with a tree," and "Dance to whatever was your favorite song exactly seven years ago." In each case, successful dancing means creatively dealing with absurd limitations—including time limits, which are designed to make the quests easy to fit into your day. It's meant to be like brushing your teeth—a little dancing every day goes a long way.

Meanwhile, the dance-offs—in which players form teams and earn points for every team member who submits a dance—require players to synchronize their efforts, even if they are dancing alone. In one of the most popular dance-offs, for example, called "Steal my bad move," players invent a signature dance move and upload a video demonstrating it. Their team gets points for every player who successfully learns and repeats the same move in their own dance-off video.

What else makes the game work? Some of the supporting design choices I made for Top Secret Dance-Off were simply twists on very traditional strategies for getting people to dance. Masks, for example, have always been an important part of persuading people to let down their guard, and play and perform. They free us from the constraints of who we think we're supposed to be and how we're supposed to behave. For people who don't see themselves as natural dancers, their TSDO disguise is meant to free them from that limiting self-identity.

**A Top Secret Dance-Off player completes
Dance Quest #1, dancing in disguise.**
(Top Secret Dance-Off by Avant Game, 2009)

But the "top secret" theme isn't just about practical considerations like obscuring player identity. It was also a lightweight way to create a kind of superhero mythology around dancing together. Dancing in front of others, after all, is an act of courage. And it's a proven powerful force for good when you inspire others to dance. Treating players like top secret superheroes just for dancing is one way to playfully recognize the meaning that dancing holds for us, and the real individual strength required to do it.

Finally, perhaps one of the most effective design elements of Top Secret Dance-Off's design isn't even about dancing specifically—it's actually an adaption of Keltner's jen ratio to the online environment. I knew that in order for

TSDO to work, players would need to feel comfortable posting potentially embarrassing videos of themselves. But on most video-sharing sites, the comments section is not exactly the kindest or friendliest place on earth. Criticism, rather than support, is the general method of feedback there, and it's often personal, ugly, and mean-spirited. So I designed the comments feature of TSDO specifically to inspire players to leave positive feedback, or none at all.

Whenever you watch another player's dance video, you have the option to reward them with a plus-one of any choreopower you want. Some choreopowers are traditional dance qualities, such as beauty, coordination, and style. Others are less traditional: humor, sneakiness, imagination, and courage. The range of choreopowers allows players to develop a unique profile of dance ability and strengths, regardless of their "natural" dance talent (or lack thereof). Perhaps my favorite choreopower is exuberance, which can be awarded to anyone who is obviously joyous and carefree.

As a result, TSDO is an environment with an off-the-charts high jen ratio. It's a place where anyone can feel safe dancing together. Indeed, more than one player has professed in the TSDO chat room that their dance quest videos were the first time anyone has seen them dance publicly in years.

Top Secret Dance-Off is a more formal hack than Cruel 2 B Kind or Tombstone Hold 'Em. There's a single, central game site, and everyone plays as part of the same online community, leveling up in the same database. But it's still an incredibly lightweight solution, from a development perspective — I launched the game within a few days of starting to design it. It's built on top of the inexpensive service Ning, which lets anyone start their own social network, much the way YouTube enabled anyone to share videos online and Blogger enabled anyone to start their own blog. There aren't fancy graphics or Flash sequences, just good mission design and community support.

I created TSDO as a happiness hack for my own life, and I hoped to play it with a few dozen friends and family members. It wound up attracting a much larger group than I'd expected. The extended social network grew to include coworkers and colleagues, acquaintances and friends of friends — all in all, about five hundred of us in total played the game together for eight weeks during its initial trial run in early 2009. (And based on its early success, a commercial version of TSDO is now in the works.)

Although TSDO can be played alone, from my observations TSDO dancing is usually at least a little bit social. Most players seem to recruit at least one partner in crime when they play, so they can film each other's dance quests and compete in the same dance-offs. And many players create group disguises for two, three, four, or even five people who plan to complete all the quests together as a single top secret unit.

Most important, TSDO helps players think of themselves as dancers—which seems to make them much more likely to dance together *in person*, when the opportunities arise. Though this isn't a scientific survey, all of my friends who have played TSDO, myself included, have found themselves dancing more often in a traditional group venue—at parties, at Bollywood dance clubs, even street festivals—long after they finished the game.

Like all of the best happiness hacks, you don't have to keep playing to maintain the benefits. A good game is *that* powerful—it can change the way you see yourself and what you're capable of forever.

WHETHER WE'RE KILLING each other with kindness, turning tombstones into full houses, or dancing in disguise, there's no way around it: sometimes we have to *sneak up* on our happiness.

Two hundred years ago, the British political philosopher John Stuart Mill suggested a subversive approach to self-help. It's an approach that has much in common with the growing community of happiness hackers. Mill argued that while happiness might be our primary goal, we can't pursue it directly. It's too tricky, too hard to pin down, too easy to scare off. So we have to set other, more concrete goals, and in the pursuit of those goals, we capture happiness as a kind of by-product. He called this approaching happiness "sideways, like a crab."[30] We can't let it know we're coming. We just kind of sneak up on it from the side.

That's exactly what happiness hacks are designed to help us do: approach happiness sideways, and as a group. In fact, with crowd games, it might be more accurate to say that hacks let us *encircle* happiness—we're all sneaking up on it from different angles together. We play these crowd games because we enjoy them in the moment and because we crave the social connectivity of a multiplayer experience. But a few intense and memorable exposures to a happiness hack can shift our ways of thinking and acting in the long run, about things as diverse as kindness to strangers, dancing, and death. And if you get enough people to shift in one place, you really can change the larger culture.

The best part about happiness hacks is that it doesn't take a lot of technological know-how or sophisticated development to create one that works. It just takes a good understanding of how games motivate, reward, and connect us. With the creativity to invent some unnecessary obstacles and the courage to playtest them with as many people as possible, anyone can dream up and share new solutions to the happiness challenges of everyday life.

Alternate reality games of all kinds are designed to make us better: happier, more creative, and more emotionally resilient. When we are better in these ways, we are able to engage with the real world more wholeheartedly—to wake up each day with a stronger sense of purpose, optimism, community, and meaning in our lives.

But big crowd games, which are the subject of the next part of this book, can do more than make us better. They can help solve some of the most urgent challenges we face as the human species.

It turns out that our ability to make ourselves better as individuals—to dive into more satisfying work, to foster real hopes of success, to strengthen our social connections, to become a part of something bigger—also helps us work together, longer, on more complex and pressing problems. Games aren't just about improving our lives today—they can help us create a positive legacy for the future.

How Very Big Games
Can Change the World

You can radically alter the nature of a game by changing the number of people playing it.

—The New Games Book[1]

How Very Big Games Can Change the World

CHAPTER ELEVEN

The Engagement Economy

On June 24, 2009, more than twenty thousand Britons joined forces online to investigate one of the biggest scandals in British parliament history—investigations that led to the resignations of dozens of parliament members and ultimately inspired sweeping political reform. How did these ordinary citizens make such a big difference? They did it by playing a game.

When the game began, the scandal had been brewing in the newspapers for weeks. According to leaked government documents, hundreds of members of parliament, or MPs, were regularly filing illegal expense claims, charging taxpayers up to tens of thousands of pounds annually for personal expenses completely unrelated to their political service. In a particularly inflammatory exposé, the *Telegraph* reported that Sir Peter Viggers, an MP from the southern coast of England, claimed £32,000 for personal gardening expenses, including £1,645 for a "floating duck island."[1]

The public was outraged and demanded a full accounting of all MP expenses. In response, the government agreed to release the complete records for four years' worth of MP claims. But in what was widely considered to be an attempt to hinder further investigation of the scandal, the government shared the data in the most unhelpful format possible: an unsorted collection of more than a million expense forms and receipts that had been scanned electronically. The files were saved as images, so that it was impossible to search or to cross-reference the claims. And much of the data had been redacted with big black blocks obscuring the detailed descriptions of items expensed. The data dump was dubbed "Blackoutgate" and called a "cover-up of massive proportions."[2]

The editors at the *Guardian* knew it would take too long for their own reporters to sort through the entire data dump and make sense of it. So they decided to enlist the public's direct help in uncovering whatever it was the authorities didn't want uncovered. In other words, they "crowdsourced" the investigation.

The term **crowdsourcing**, coined by technology journalist Jeff Howe in 2006, is shorthand for outsourcing a job to the crowd.[3] It means inviting a large group of people, usually on the Internet, to cooperatively tackle a big project. Wikipedia, the collaboratively authored online encyclopedia created by a crowd of more than 10 million unpaid (and often anonymous) writers and editors, is a prime example. Crowdsourcing is a way to do collectively, faster, better, and more cheaply what might otherwise be impossible for a single organization to do alone.

With a million uncataloged government documents on its hands and no way of knowing which document could prove to be the smoking gun for which MP, the *Guardian* knew it needed all the crowd help it could get. So it decided to tap into the wisdom of the crowds—not with a wiki, however, but with a game.

To develop the game, they turned to a young, but accomplished, London-based software developer named Simon Willison. His task: convert and condense all the scanned forms and expenses into 458,832 online documents, and set up a website where anyone could examine the public records for incriminating details. For the price of just a week's worth of the development team's time and a paltry fifty pounds to rent temporary servers to host the documents, the *Guardian* launched Investigate Your MP's Expenses, the world's first massively multiplayer investigative journalism project.

⟳ HOW TO PLAY INVESTIGATE YOUR MP'S EXPENSES

Join us in digging through the MPs' expenses to review each document. Your mission: Decide whether it contains interesting information, and extract the key facts.

Some pages will be covering letters or claim forms for office stationery. These can be safely ignored.

But somewhere in here is the receipt for a duck island. And who knows what else may turn up. If you find something which you think needs further attention, simply hit the button marked "Investigate this!" and we'll take a closer look.

Step 1: Find a document.

Step 2: Decide what kind of thing it is (expenses claim, proof/ receipt, or blank)

Step 3: Transcribe the line items

Step 4: Make any specific observations about why a claim deserves further scrutiny

Examples of things to look out for: food bills, repeated claims for less than £250 (the limit for claims not backed up by a receipt), and rejected claims.

Investigate your own MP: Enter your postal code to bring up all of your MP's claims and receipts. Or investigate by political party.

All the MPs' records are on there now—so let us know what you find.

Just three days into the game, it was clear that the crowdsourcing effort was an unprecedented success. More than 20,000 players had already analyzed more than 170,000 electronic documents. Michael Andersen, a member of the Nieman Journalism Lab at Harvard University and an expert on Internet journalism, reported at the time: "Journalism has been crowdsourced before, but it's the scale of the *Guardian*'s project—170,000 documents reviewed in the first 80 hours, thanks to a visitor participation rate of 56 percent—that's breathtaking."[4]

A visitor participation rate measures the percentage of visitors who sign up and make a contribution to a network. A rate of 56 percent for *any* crowdsourced project was unheard of previously. (By comparison, roughly 4.6 percent of visitors to Wikipedia make a contribution to the online encyclopedia.)[5] It's *especially* breathtaking considering the mind-numbingly tedious nature of the actual accounting work being performed.

So what accounted for this unprecedented participation in a citizen journalism project? According to Willison, it all boiled down to rewarding participants in the right way: with the emotional rewards of a good game.

"The number one lesson from this project: Make it feel like a game," Willison said in an interview with the Nieman Journalism Lab. "Any time that you're trying to get people to give you stuff, to do stuff for you, the most important thing is that people know that what they're doing is having an effect. If you're not giving people the 'I rock' vibe, you're not getting people to stick around."

The "I rock" vibe is another way of talking about classic game rewards, such as having a clear sense of purpose, making an obvious impact, making continuous progress, enjoying a good chance of success, and experiencing plenty of fiero moments. The Investigate Your MP's Expenses project featured all of these emotional rewards, in droves.

The game interface made it easy to take action and see your impact right away. When you examined a document, you had a panel of bright, shiny buttons to press depending on what you'd found. First, you'd decide what kind of document you were looking at: a claim form, proof (a receipt, invoice, or purchase order), a blank page, or "something we haven't thought of." Then you'd determine the level of interest of the document: "Interesting," "Not interesting," or "Investigate this! I want to know more." When you'd made your selection, the button lit up, giving you a satisfying feeling of productivity, even if all you'd found was a blank page that wasn't very interesting. And there was always a real hope of success: the promise of finding the next "duck pond" to keep you working quickly through the flow of documents.

A real-time activity feed showed the names of players logged in recently and the actions they'd taken in the game. This feed made the site feel social. Even though you were not directly interacting with other players, you were copresent with them on the site and sharing the same experience. There was also a series of top contributor lists, for the previous forty-eight hours as well as for all time, to motivate both short-term and long-term participation. And to celebrate successful participation, as well as sheer volume of participation, there was also a "best individual discoveries" page that identified key findings from individual players. Some of these discoveries were over-the-top luxuries offensive to one's sense of propriety: a £240 giraffe print or a £225 fountain pen, for example. Others were mathematical errors or inconsistencies suggesting individuals were reimbursed more than they were owed. As one player noted, "Bad math on page 29 of an invoice from MP Denis MacShane, who claimed £1,730 worth of reimbursement, when the sum of those items listed was only £1,480."

But perhaps most importantly, the website also featured a section labeled "Data: What we've learned from your work so far." This page put the individual players' efforts into a much bigger context—and guaranteed that contributors would see the real results of their efforts. Some of the key results of the game included these findings:

- On average, each MP expensed *twice* his or her annual salary, or more than £140,000 in expenses on top of a £60,675 salary.
- The total cost to taxpayers of personal items expensed by MPs is £88 million annually.

And the game detailed:

- The number of receipts and papers filed by each MP, ranging between 40 and 2,000

- The total expense spending by party and by category (kitchen, garden, TV, food, etc.)
- Online maps comparing travel expenses filed with actual distance from the House of Commons in London to the MPs' home districts, making it easy to spot MPs grossly overcharging for travel (for example, MPs from nearby districts who filed £21,534 versus £4,418, or £10,105 versus £1,680)

Bringing these numbers to light helped clarify the true extent of the crisis: a far more pervasive culture of extravagant personal reimbursement than originally suspected.

So what did the players accomplish? Real political results. At least twenty-eight MPs resigned or declared their intention to do so at the end of their term, and by early 2010 criminal proceedings against four MPs investigated by the players were under way. New expense codes are being written, and old codes are being enforced more vigorously. Most concretely, hundreds of MPs were ordered to repay a total of £1.12 million.[6]

It's not all the doing of the *Guardian*'s gamers, of course. But without a doubt, the game played a crucial role. The citizen journalists helped put significant political pressure on the British government by keeping the scandal in the news. The longer the game continued, the more public momentum built to force major policy reform.

Investigate Your MP's Expenses enabled tens of thousands of citizens to participate directly in a new kind of political reform movement. Instead of just clamoring for change, they put their time and effort into creating evidence that change was needed. Crucially, the crowd of gamers did all of this important work faster than any individual organization could have, and they did it for free—lowering the costs of investigative journalism and speeding up the democratic reform process.

Not all crowdsourcing projects are so successful. Working together on extreme scales is easier said than done. You can't crowdsource without a crowd—and it turns out that actively engaged crowds can be hard to come by.

In 2008, New York University professor and Internet researcher Clay Shirky sat down with IBM researcher Martin Wattenberg and tried to work out exactly how much human effort has gone into making Wikipedia. They looked at the total number of articles and edits, as well as the average article length and average time per edit. They factored in all of the reading time required to find knowledge gaps and spot errors, and all of the hours of programming and ongoing community management required to make those edits hang together coherently. After a lot of clever math, they worked out the following estimate:

> If you take Wikipedia as a kind of unit, all of Wikipedia, the whole project—every page, every edit, every talk page, every line of code, in every language that Wikipedia exists in—that represents something like the accumulation of 100 million hours of human thought. . . . It's a back-of-the-envelope calculation, but it's the right order of magnitude, about 100 million hours of thought.[7]

On one hand, that's no trivial effort. It's the equivalent of rounding up a million people and convincing them to spend a hundred hours each contributing to Wikipedia, for free. Put another way: it's like persuading ten thousand people to dedicate five full-time work years to the Wikipedia project. That's a *lot* of effort to ask a *lot* of people to make, for no extrinsic reward, on behalf of someone else's vision.

On the other hand, given that there are 1.7 billion Internet users on the planet and twenty-four hours in a day, it really shouldn't be that hard to successfully pull off lots of projects on the scale of Wikipedia.[8] Hypothetically, if we could provide the right motivation, we should be able to complete one hundred Wikipedia-size projects *every single day*—if we could convince all 1.7 billion Internet users to spend most of their free time voluntarily contributing to crowdsourced projects.

Maybe that's unrealistic. More reasonably, if we could convince every Internet user to volunteer just one single hour a week, we could accomplish a great deal. Collectively, we would be able to complete nearly *twenty* Wikipedia-size projects *every single week*.

Which really makes you wonder: with so much potential, why aren't there even more Wikipedia-scale projects out there?

The truth is, the Internet is littered with underperforming, barely populated, or completely abandoned collaboration spaces: wikis that have no contributors, discussion forums with no comments, open-source projects with no active users, social networks with barely a few members, and Facebook groups with plenty of members but few who ever do anything after joining. According to Shirky, more than half of all collaborative projects online fail to achieve the minimum number of participants necessary to even begin working on their goal, let alone achieve it.

It's not for a lack of time spent on the Internet. It's just incredibly difficult to achieve the necessary critical mass of participation on any given serious project.

For one thing, some participatory networks are more rewarding than others— and the most readily rewarding networks aren't, as a rule, the ones doing serious work. Online games and "fun" social networks like Facebook provide the steadiest stream of intrinsic rewards. They're *autotelic* spaces—spaces we visit for the pure enjoyment of it. Their primary purpose is to be rewarding, not to solve a problem or get work done. Unlike serious projects, they are engineered first and foremost to engage and satisfy our emotional cravings. And as a result, *they* are the projects that are absorbing the vast majority of our online **participation bandwidth**—our individual and collective capacity to contribute to one or more participatory networks.

A second and more pressing problem is the fact that, across serious crowd projects, our participation resources are increasingly being spread too thin.

In the past month, I've been invited to join exactly forty-three Facebook groups. I've been asked to help edit fifteen wikis and contribute to nearly twenty Google Docs. And I've been (unsuccessfully) recruited for nearly twenty other assorted collective intelligence projects, each one requesting me to spend valuable online time voting, ranking, judging, editing, sorting, labeling, approving, commenting, translating, predicting, contributing, or otherwise participating in someone else's idea of a worthy mission. I may be an extreme example—I'm a highly networked individual with many personal contacts doing interesting work online. But I'm certainly not alone in feeling overwhelmed by participation requests. Increasingly, I hear the same complaint from friends, colleagues, and clients: there are simply too many demands, from too many people, on our online engagement.

I call it "participation spam." It's the increasingly unsolicited requests we receive on a daily basis to participate in someone else's group. If you're not getting participation-spammed yet, you will—and soon.

By my own back-of-the-envelope estimate, there are currently more than 200 million public requests for crowd participation on the Internet, across thousands of different networks, ranging from citizen journalism, citizen science, and open government to peer-to-peer advice, social networking, and open innovation. This estimate factors in, for example, more than 1 million public social networks created on Ning, more than 100,000 wikis on Wikia, more than 100,000 crowdsourcing projects on Amazon's Mechanical Turk, at least 20,000 videos awaiting transcription and translation on DotSUB, as well as myriad smaller clusters of open collaboration, such as the more than 3,300 public "idea spaces" for proposing and developing innovative ideas on IBM Lotus' IdeaJam and more than 14,000 on Dell's IdeaStorm.

With 1.7 billion people on the Internet, that works out to about 8.5 people per crowd.

That's a very small crowd.

It's certainly not a big enough crowd to build a resource on the scale of Wikipedia.

This problem is likely going to get worse before it gets better. As it becomes easier and cheaper to launch a participation network, it will likely become equally difficult to sustain it. There are only so many potential participants on the Internet. And as long as participation is designed as an active process requiring some mental effort, there are only so many units of engagement, or mental hours, each participant can reasonably expend in a given hour, day, week, or month.

To effectively harness the wisdom of the crowds, and to successfully leverage the participation of the many, organizations will need to become effective players in an emerging **engagement economy**. In the economy of engagement, it is less and less important to compete for attention and more and more important to compete for things like brain cycles and interactive bandwidth. Crowd-dependent projects must figure out how to capture the mental energy and the active effort it takes to make individual contributions to a larger whole. For this reason, the overall crowdsourcing culture likely will not be immune from "the tragedy of the commons"—the crisis that occurs when individuals selfishly exhaust a collective resource. Collaboration projects will have to compete for crowd resources as online communities seek to grab as many mental hours as possible from their members. These gains will likely come at the expense of other projects still striving to secure their own passionate community. Collaboration may be the signature modus operandi of these projects, but the competition for participants will be fierce and not all projects will thrive.

As we consider these challenges, some of the key questions for the emerging engagement economy start to arise: Who will do all of the participating necessary to make the seemingly endless flow of participatory projects a success? Are there enough willing quality collaborators in the world to do it? How do you draw a big and passionate enough crowd to tackle extreme-scale goals? And what will motivate the crowds who do show up to stick around long enough to collectively create something of value?

We have to face facts. It's very difficult to motivate large numbers of people to come together at the same time and to contribute any significant amount of energy—let alone their very best effort—to a collaborative project. Most big crowd projects today fail: they fail to attract a crowd, or they fail to give the crowd the right kind of work, or they fail to reward the crowd well enough to keep it participating over the long haul.

But it's not hopeless. As both Wikipedia and Investigate Your MP's Expenses show, there are significant crowdsourcing projects succeeding. And they all have one important thing in common: they're structured like a good multiplayer game.

The most active contributors to Wikipedia, the world's most successful crowdsourced project, already know this. In fact, they've created a special project to detail all the ways in which Wikipedia is like a game.

As more than fifty leading Wikipedia contributors have helped explain,

"One theory that explains the addictive quality of Wikipedia and its tendency to produce Wikipediholics (people who are addicted to editing Wikipedia articles) is that Wikipedia is a massively multiplayer online role-playing game." And, according to the happily addicted Wikipedians, it works like a good MMORPG in three key ways.

First, Wikipedia is a **good game world**. Its extreme scale inspires our sense of awe and wonder, while its sprawling navigation encourages curiosity, exploration, and collaboration. Here's exactly how the Wikipedians described it in one recent version of the constantly changing wiki page:

> Wikipedia has an immersive game world with over 10.7 million *players* (registered contributors, or "Wikipedians") and over 3.06 million *unique locations* (Wikipedia articles), including 137,356 undiscovered *secret areas* ("lonely pages," or articles that aren't linked to any other articles and therefore can't be found by browsing), 7,500 completely explored *dungeons* ("good articles," or exhaustively written articles with excellent citations and evidence provided), and 2,700 *boss levels* ("featured articles," or the top-ranking articles as judged by accuracy, neutrality, completeness, and style).[9]

In other words, Wikipedia, like all of the most engaging multiplayer game worlds, is an *epic built environment*. It invites participants to explore, act, and spend large amounts of time there.

Second, Wikipedia has **good game mechanics**. Player action has direct and clear results: edits appear instantly on the site, giving users a powerful sense of control over the environment. This instant impact creates optimism and a strong sense of self-efficacy. It features unlimited work opportunities, of escalating difficulty. As the Wikipedians describe it, "Players can take on *quests* (WikiProjects, efforts to organize many articles into a single, larger article), fight *boss-level battles* (featured articles that are held to higher standards than ordinary articles), and enter *battle arenas* (interventions against article vandalism)." It also has a personal feedback system that helps Wikipedians feel like they are improving and making personal progress as they contribute. "Players can accumulate *experience points* (edit count), allowing them to advance to higher *levels* (lists of top-ranking Wikipedians by number of edits)."

Meanwhile, like all good games, there is significant friction to achieving the goal. It's not just about making good edits. The game also has a clearly defined enemy to defeat: vandals who make unhelpful edits to the site. "Edit wars" are said to break out between competing contributors with different points of view, and players have developed collaboration techniques and combat tools to deal with these high-level challenges. As an edit war escalates, more and more editors are called to join the conversation and work toward a solution.

Which leads to the third key aspect of Wikipedia's good gameness: it has **good game community**. Good game community requires two things: plenty of positive social interaction and a meaningful context for collective effort. Wikipedia has both. As Wikipedians describe it:

> Every *unique location* (article) in the *game world* (encyclopedia) has a *tavern* ("talk page," or discussion forum) where players have the opportunity to interact with any other player in real time. Players often become friends with other players, and some have even arranged to *meet in real life* ("meetups," or face-to-face social gatherings for frequent Wikipedia contributors).

The talk pages encourage both sociable competition (arguing over recent edits) and collaboration (improving and organizing existing articles). This kind of persistent positive social interaction around common goals builds trust and strong bonds—which naturally extends to face-to-face relationships. Indeed, roughly a hundred Wikipedia meetups occur a year, everywhere from Reykjavik, Cape Town, Munich, and Buenos Aires to Perth, Kyoto, Jakarta, and Nashville.

Also crucial to good community is the sense of meaning created by belonging and contributing to such an epic project. Wikipedia members are always working toward extreme-scale goals—aiming first for 100,000 articles, then 1 million, then 2 million, and then 3 million—as well as celebrating traffic milestones—the date Wikipedia first broke into the top 500 websites, the top 100, the top 20, and, most recently, the top 10. And they are constantly immersed in awe-inspiring project statistics, greeted on the site's home page with a list of the more than 270 different language versions of Wikipedia and growing.

Wikipedians explicitly credit the good gameness of the system—its compelling game world, satisfying game mechanics, and inspiring game community—for their dedicated long-term participation. To conclude their analysis of Wikipedia as an MMORPG: "People tend to play a given MMORPG for six to eighteen months at a high level of involvement; a similar pattern (of "Wiki-breaking," or separating from the site to attend to other projects) has been noted in hard-core Wikipedia players."[10] In other words, most games eventually get boring—we exhaust their challenges and creative possibilities—and Wikipedia is no different. While there are some perpetual Wikipedians, most members are of service to the site for a limited period of time, after which they're likely to move on to a new system that offers new content and fresher challenges.

The "Wikipedia is an MMORPG" project is particularly compelling precisely because so much valuable participation effort is being spent in MMORPG environments.

Take *World of Warcraft*, for example—the most successful MMORPG ever. Currently, with more than 11.5 million subscribers, each averaging between sixteen and twenty-two hours a week playing the game, that's 210 million participation hours spent weekly on just a single MMORPG. And the number of WoW subscribers is almost exactly the same as the number of registered contributors to Wikipedia.

Based on Clay Shirky's estimate that all of Wikipedia took 100 million hours to create, the WoW community alone could conceivably create a new Wikipedia every three and a half days.

But let's say, for argument's sake, that most people who play WoW wouldn't be even remotely interested in any kind of collective intelligence project. There are still more than 65,000 WoW players who are registered contributors to WoWWiki. Even if you managed to successfully engage only that group, it would still take them only *two months* of channeling their usual WoW playing time to a crowdsourcing project to collectively create a resource on the scale of Wikipedia. By comparison, Wikipedia took *eight years* to collect 100 million hours of cognitive effort.

When I first started looking at these numbers, I had two insights.

First, gamers are an extremely valuable—and largely untapped—source of participation bandwidth. Whoever figures out how to effectively engage them first for real work is going to reap enormous benefits. (And clearly, the Guardian's Investigate Your MP's Expenses represents one of the first organizations to do just that.)

Second, crowdsourcing projects—if they have any hope of capturing enough participation bandwidth to achieve truly ambitious goals—must be intentionally designed to offer the same kinds of intrinsic rewards we get from good games. Increasingly, I'm convinced that this is the only way to dramatically increase our total available participation bandwidth. If everyone spent as much time actively engaged in good, hard work as gamers do, we wouldn't be competing for scarce crowd resources. We'd have massively more mental hours to pour into important collective efforts.

Making Better Use of Gamers' Participation Bandwidth

My experience and research suggests that gamers are more likely than anyone on the planet to contribute to an online crowdsourcing project. They already have the time and the desire to tackle voluntary obstacles. They're playing games precisely because they hunger for more and better engagement. They also have proven computer skills and an ability to learn new interactive interfaces quickly. And if they're playing games online, they already have the necessary network access to join any online project and start participating immediately.

Given the highly social nature of today's best games, gamers are also very likely to have a large network of friends and family they already bring from one game to the next. This is exactly the kind of social infrastructure necessary to help grow any participation base.

On the whole, gamers already spend more time compiling collective intelligence—and making effective use of it—than anyone else. They're the most prolific users of wikis in the world. On Wikia, for example, the most popular wiki-hosting service, gamers are by far the leading creators of content and the most active users. With more than a million articles on ten thousand distinct wikis—each wiki for a different game—they represent the lion's share of active content across the entire Wikia network. And as Artur Bergman, vice president of engineering and operations at Wikia, has told me many times, they are by far the most organized and ambitious wiki users on the network. "The gamers are amazing," he said this fall, after watching multiple game walk-throughs go up overnight for newly released games. "The minute the game comes out, they start making round-the-clock edits. Within twenty-four hours, they have the whole thing documented."

The minute gamers get their hands on a new game, they start compiling collective intelligence about it. It's not something that happens after they get tired of playing—it's an essential part of gaming. And, according to Wikia's traffic statistics, for every single wiki contributor, thousands more players show up to make use of the data. Gamers make daily use of collective intelligence, and as a result they instinctively understand the value and possibility of big crowd projects.

In short, gamers are already our most readily engageable citizens.

We also have ample proof that gamers want to do more than just save the virtual world. Two key projects show just how much online gamers want to do real-world good: the world hunger–fighting game Free Rice and the cancer-fighting gamer initiative Folding@home on the PlayStation 3.

FREE RICE—OR HOW TO PLAY AND FEED HUNGRY PEOPLE

"Feeling guilty about wasting time on computer solitaire? Join the growing guilt-free multitude at FreeRice.com, an online game with redeeming social value."[11] That's how *USA Today* described Free Rice, a nonprofit game designed to help gamers battle world hunger while they play.

The gameplay is simple: answer a multiple-choice vocabulary question correctly, and you earn ten virtual grains of rice. The better you do, the harder the questions get; it took me only six questions in my latest game before getting stumped by this one:

> *Acrogenous* **means:**
> - created top down
> - extremely generous
> - growing from the tip of a stem
> - pointy-headed

(*Hint*: It turns out "acrogenous" is a botanical term—see the endnotes for the answer.)[12]

You can stack up as much virtual rice as you want, and at the end of your game, it gets converted to real rice, which is donated to the United Nations World Food Programme. (The rice is provided by sponsors whose online advertisements appear underneath every question in the game.)

To earn enough rice to feed one person one meal, I'd have to answer two hundred questions correctly. But it's not the kind of game you really want to play for hours on end. In fact, usually I just play for about a minute or two, or roughly ten questions at a time, whenever I want a quick burst of satisfying productivity and feel-good activity. But earning a hundred grains a day is barely a teaspoon's worth; luckily I'm not the only person playing. On any given day, between two hundred thousand and five hundred thousand people play Free Rice; together, according to the game's FAQ, their efforts add up to enough rice to feed an average of seven thousand people per day.

Why is Free Rice able to capture so much engagement? It isn't just that it is a force for good; it's also classically good game design. It takes just seconds to complete a task, meaning you can get a lot of work done quickly. You get instant visual feedback: grains of rice stacking up in a bowl, with a constantly rising total of grains that you've earned. Because the game gets easier when you make mistakes and harder when you answer correctly, it's easy to experience flow: you're always playing at the limits of your ability. And since the game was created, in 2007, its game world has expanded significantly: there's a seemingly endless stream of potential tasks, with thirteen different subject areas, from famous paintings and world capitals to chemical symbols and French vocabulary. There's also a clear sense that you're a part of something bigger as you play. As the Free Rice site explains, "Though 10 grains of rice may seem like a small amount, it is important to remember that while you are playing, so are thousands of other people at the same time. It is everyone together that makes the difference."[13] So far, that difference is nothing less than epic: 69,024,128,710 grains of rice and counting—enough to provide more than 10 million meals worldwide.

Free Rice in one respect seems like a perfect embodiment of the crowdsourcing philosophy: lots of people come together to make a small contribution, all of it adding up to something bigger. But Free Rice actually falls short of real crowdsourcing. That's because the grains of rice aren't coming from the players—they're coming from a small number of advertisers who agree to pay the cost of ten grains of bulk rice for every correct-answer page view. Those advertisers are paying for the gamers' eyeballs on the page. So the actual gameplay activity isn't generating any new knowledge or value; the advertisers are just happy to have their advertisement on a page they know hundreds of thousands of people will see daily.

That means Free Rice is less like Wikipedia and more like clever fundraising. But Free Rice is still an extremely important project, for one big reason: it irrefutably shows that gamers are, on the whole, happier when a good game also does real-world good. There's no evidence that hundreds of thousands of people would show up to play a bad game just to help out a good cause. But the combination of good game design and real-world results is irresistible.

It also points the way to bigger possibilities. What if people playing Free

Rice were actually contributing something other than their attention to advertising? What could gamers easily contribute, and what would it add up to? We can catch an even better glimpse of gamers' potential to engage in epic problem solving in a different crowd project: Folding@home, a project designed to harness gamers' hardware for good.

FOLDING@HOME ON PLAYSTATION 3

"If you own a PS3, start saving lives. *Real* lives."[14] That's how one blogger put it when he discovered Folding@home for the PlayStation 3, the world's first distributed computing initiative just for gamers. A distributed computing system is like crowdsourcing for computers. It connects individual computers via the Internet into a giant virtual supercomputer in order to tackle complex computational tasks that no individual computer could solve alone.

For years, scientists have been harnessing the processing power of home computers to create virtual supercomputers tasked with solving real scientific problems. The most famous example is SETI@home, or Search for Extraterrestrial Intelligence at home, a program that harnesses home computers to analyze radio signals from space for signs of intelligent life in the universe. Folding@home is a similar system created by biologists and medical researchers at Stanford University in an effort to solve one of the greatest mysteries of human biology: how proteins fold.

Why is protein folding important? Proteins are the building blocks of all biological activity. Everything that happens in our bodies is a result of proteins at work: they support our skeleton, move our muscles, control our five senses, digest our food, defend against infections, and help our brain process emotions. There are more than one hundred thousand kinds of proteins in the human body, each consisting of anywhere from one hundred to one thousand different parts and made up of any combination of twenty different amino acids. In order to do its specific job, each kind of protein folds up into a unique shape.[15]

Biologists describe this process as a kind of incredibly complex origami.

The parts can be arranged and folded up in almost any imaginable combination and form. Even if you know which amino acids make up a protein, and in how many parts, it's still nearly impossible to predict exactly what form the protein will take. One thing scientists know for sure, however, is that sometimes, for unknown reasons, proteins stop folding correctly. They "forget" what shape to take—and when they do, this can lead to disease. Alzheimer's disease, cystic fibrosis, Mad Cow disease, and even many cancers, for example, are believed to result from protein misfolding.

So scientists want to understand exactly how proteins fold and what shapes they take, in order to figure out how to stop proteins from misfolding. But given the nearly infinitely many different shapes each protein can take, it requires an incredibly long time to test all the various potential shapes. Computer programs can simulate every possible shape that a protein with a certain amino acid composition could make. But it would take *thirty years* to test all the different combinations for just one single protein, out of the hundred thousand proteins in our bodies. As the Folding@home FAQ section puts it, "That's a long time to wait for one result!"

That's why scientists use distributed computing systems. By dividing the work between multiple processors, the work can go much, much faster. Since 2001, anyone in the world has been able to connect their personal computer to the Folding@home network. Whenever their computer is idle, it connects to the network and downloads a small processing assignment—just a few minutes' worth of protein-folding simulation. It submits the data to the network when it's done.

But after nearly a decade of tapping into the spare processing power of personal computers, the team behind Folding@home realized that a more powerful platform for virtual supercomputing exists: game consoles like the PlayStation 3.

When it comes to data-crunching ability, game consoles are significantly more powerful than the average PC. That's because the computational power required to render constantly changing 3D graphic environments is much greater than what's required for ordinary home or work computing tasks, like

Internet browsing or word processing. Even though there are collectively many more PCs in our homes than game consoles, if scientists could get even a small fraction of gamers to participate in distributed computing projects, they could double, triple, or even quadruple their supercomputing power.

But would gamers do it? Sony, the makers of the PS3 console, bet that they would. And they were right.

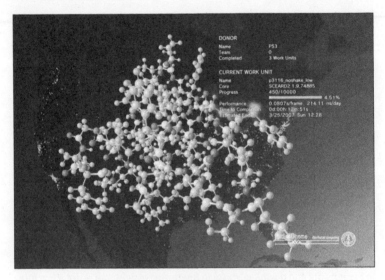

Screenshot of the Folding@home application
for the PlayStation 3 system.
(Sony Corporation, 2009)

As a philanthropic venture, Sony developed a custom Folding@home application for the PS3. Gamers could log in, accept a protein-folding mission, and donate the power of their PS3 to get the mission done. They could watch the folding simulation in action, and keep tabs on just how much computational effort they'd personally contributed to the project.

Help save real lives when you're not saving virtual lives. The message was compelling, and it caught on fast. Within days of Sony releasing the application, thousands of blog posts and online articles about "gaming for the greater good" spread across the entire online gamer landscape.

The gamer community has rallied around the mission with enormous enthusiasm. Online articles and blog posts proudly proclaimed: "PS3 Gamers Trying to Save the World!" On forums, players encouraged each other: "Have you cured cancer lately? Now is the best time to jump in and join the cancer-saving fun."[16] They set up competitive folding teams and tried to rally each other to action: "Your PS3 can't do it without you." "It's time to do your part for humanity."[17]

Within six months, gamers collectively helped the Folding@home network achieve supercomputing milestones never achieved by any other distributed computing network anywhere in the world. As a senior developer for the PS3 Folding@home project announced on the official PlayStation blog:

> This time it's something that the Folding community and the computer science field as a whole have been anxiously awaiting—the crossing of a milestone known as a petaflop. A petaflop equals one quadrillion floating point operations per second (FLOPS). If you'd like to imagine this enormous computation capacity, think about calculating a tip on a restaurant bill, now do that for 75,000 different bills, now do that every second, and lastly, imagine everybody on the planet is doing those calculations at the same second—this is a petaflop calculation. Now you see why I say enormous . . .[18]

With an epic context like that, it's no wonder gamers rose to the occasion. They live for opportunities to be of service to extreme-scale goals. As one gamer said, "You might as well be bragging that you helped cure cancer, instead of just beating the game on the hardest difficulty level without dying once."[19]

Today, PS3 users account for 74 percent of the processing power used by Folding@home. So far, more than a million PS3 gamers on six continents have contributed spare computing cycles to the Folding@home project. That's one out of every twenty-five gamers on the PS3 network.[20] The gamers are vastly outcontributing everyone else on the network—and they're far more active on Folding@home forums, keeping close tabs on what their efforts are adding up to.

Now every PS3 comes preloaded with Folding@home software, making it even easier for any gamer to opt in to a scientific mission. As it stands, long after the initial September 2008 launch, gamers continue to sign up for the collective effort at the rate of three thousand a day, or two new volunteers every minute.[21]

The Folding@home project for the PlayStation 3 is a perfect example of matching ability with opportunity, which is the fundamental dynamic of any good crowdsourcing project. It's not enough to draw a crowd—you have to ask the crowd to do something they have a real chance of doing successfully. Every PS3 gamer is capable of easily and successfully contributing spare processing power. Meanwhile, Sony, working with Stanford University, has created an opportunity for that contribution to really mean something.

GAMERS' MASS PARTICIPATION in, and enthusiasm for, this big crowd project is a clear sign that there is a growing desire to be of service to real-world causes. For decades, gamers have been answering heroic calls to action in virtual worlds. It's time we ask them to answer real-world calls to action, and all the evidence suggests that they are more than happy—they are *happier*—to rise to the real-world occasion.

The next major step to take, then, will be to harness gamers' minds, and not just their consoles. Gamers are creative, persistent, and always up for a good challenge. Their strong cognitive resources, combined with their proven engageability, are a valuable resource just waiting to be tapped. In fact, a team of medical scientists, computer scientists, engineers, and professional game developers from the Seattle area are banking on that fact. They believe that gamers can use their natural creative ability and problem-solving abilities to learn to design new protein shapes and actively help cure diseases. They've created a protein-folding game called Foldit, which represents a dramatic leap of faith forward from the Folding@home project.

Instead of harnessing their video game *hardware* to run complex protein-folding simulations, Foldit harnesses the real brain power of gamers, challenging them to use their creativity and ingenuity to fold digital proteins by hand.

In the game, players manipulate proteins in a 3D virtual environment that one reviewer describes as "a twenty-first-century version of Tetris, with multicolored geometric snakes filling the screen."[22] The geometric snakes represent all the different building blocks of a protein, the amino acid chains that connect and fold up into incredibly complex patterns in order to perform different biological tasks in the body. In Foldit, the player's goal is to learn what kinds of patterns are the most stable and successful for doing different jobs, by taking an unfolded protein and folding it up into the right shape. This is called a "protein puzzle."

Players learn how to fold proteins by working on "solved" puzzles, or proteins that scientists already know how to fold. Once they've got the hang of it, they're encouraged to try to predict the shape of a protein that scientists haven't successfully folded yet, or to design a new protein shape from scratch, which researchers could then manufacture in a lab.

"Our ultimate goal is to have ordinary people play the game and eventually be candidates for winning the Nobel Prize in biology, chemistry or medicine." Zoran Popović, a professor of computer science and engineering at the University of Washington, and one of the lead researchers on the Foldit project, declared these Nobel aspirations in his address to the Games for Health conference in the spring of 2008, just weeks before opening up the new protein-folding puzzle game to the public.[23] Within eighteen months of its release, the game had attracted a registered community of more than 112,700 players—most of whom, according to researchers, had little to no previous experience in the field of protein folding. "We're hopefully going to change the way science is done, and who it's done by," Popović said.

The Foldit team is well on its way to doing just that. In the August 2010 issue of the prestigious scientific journal *Nature*, the team declared its first significant breakthrough: In a series of ten challenges, gamers beat the world's most sophisticated protein-folding algorithms five times, and drew even three times. The authors concluded that gamer intuition can successfully compete with supercomputers—especially when the problems being solved require taking radical, creative risks.[24] Most notably, the *Nature* study wasn't just *about* the Foldit players; it was *by* the Foldit players. More than 57,000 gamers were listed as official coauthors alongside Popović and his university colleagues.[25]

IN THE DECADES to come, there will be many more challenges for us to tackle together as crowds: more citizen journalism investigations, more collective intelligence projects, more humanitarian efforts, more citizen science research. There's no shortage of world-changing collective work to be done—so we can't allow ourselves to be limited by a shortage of incentive or compensation.

Many crowdsourcing projects today are experimenting with micropayments, or small amounts of monetary reward, in return for contributions. The Amazon Mechanical Turk marketplace, which gives businesses access to a global virtual workforce, pays participants a few cents for each helpful contribution to a **human intelligence task**, or HIT—a cognitive task "that only a human, and not a computer, could do" (such as labeling images, characterizing the emotional quality of song lyrics, or describing the action in short videos). Others offer prizes for top contributions. The CrowdSPRING marketplace, for example, offers prizes starting at $5,000 to individuals who submit the most helpful ideas—for example, helping name a new product or improve an existing service.

The logic behind these practices is that if people are willing to contribute for free, they'll be even happier to contribute when they're compensated. But compensating people for their contributions is *not* a good way to increase global participation bandwidth, for two key reasons.

First, as numerous scientific studies have shown, compensation typically *decreases* motivation to engage in activities we would otherwise freely enjoy.[26] If we are paid to do something we would otherwise have done out of interest—such as reading, drawing, participating in a survey, or solving puzzles—we are less likely to do it in the future without being paid. Compensation increases participation only among groups who would never engage otherwise—and as soon as you stop paying them, they stop participating.

Second, there are natural limits on the monetary resources we can provide a community of participants. Any given project will have only so much financial capital to give away; even a successful business will eventually hit an upper limit of what it can afford to pay for contributions. Scarce rewards like money and prizes artificially limit the amount of participation a network can inspire and support.

We need a more **sustainable engagement economy**—an economy that works by motivating and rewarding participants with intrinsic rewards, and not more lucrative compensation.

So if not money or prizes, then what will most likely emerge as the most powerful currency in the crowdsourcing economy? I believe that emotions will drive this new economy. Positive emotions are the ultimate reward for participation. And we are already hardwired to produce all the rewards we could ever want—through positive activity, positive achievements, and positive relationships. It's an infinitely renewable source of incentive to participate in big crowd projects.

In the engagement economy, we're not competing for "eyeballs" or "mindshare." We're competing for brain cycles and heartshare. That's why success in the new engagement economy won't come from providing better or more competitive compensation. It will come from providing better and more *competitive engagement*—the kind of engagement that increases our personal and collective participation bandwidth by motivating us to do more, for longer, toward collective ends. And no one knows how to augment our collective capacity for engagement better than game developers.

Game designers have been honing the art of mass collaboration for years. Games inspire extreme effort. Games create communities that stick together over time, long enough to get amazing things done together. If crowdsourcing is the theory, then games are the platform.

Which brings us to our next fix for reality:

⟳ FIX #11: A SUSTAINABLE ENGAGEMENT ECONOMY

Compared with games, reality is unsustainable. The gratifications we get from playing games are an infinitely renewable resource.

Good game developers know that the emotional experience itself is the reward. Consider the following job listing for Bungie, the company that creates the *Halo* video game series:

> Do you dream about creating worlds imbued with real value and consequence? Can you find the fine line between a reward that encourages players to have fun and an incentive that enslaves them? Can you devise a way for a player to grow while preserving a delicate game balance? If you answered yes to these questions, you might want to polish up your résumé and apply to be Bungie's next Player Investment Design Lead.
>
> The Player Investment Design Lead directs a group of designers responsible for founding a robust and rewarding investment path, supported by consistent, rich and secure incentives that drive player behavior toward having fun and investing in their characters, and then validates those systems through intense simulation, testing and iteration.[27]

This kind of job doesn't yet exist outside of the game industry. But it should. "Player investment design lead" is a role that every single collaborative project or crowd initiative should fill in the future. When the game is intrinsically rewarding to play, you don't have to pay people to participate—with real currency, virtual currency, or any other kind of scarce reward. Participation is its own reward, when the player is properly invested in his or her progress, in exploring the world fully, and in the community's success.

So how exactly do you design good player investment? The Bungie job listing further details some of the core responsibilities of the position—and, in a nutshell, they give us a very good idea of four engagement principles any big crowd project should follow. As you can see, these four principles all serve the ultimate goal of building a compelling game world, satisfying game mechanics, and an inspiring game community.

The Player Investment Design Lead will design the mechanics that drive in-game player reward and incentives:
- So players feel invested in the world and their character.
- So players have long-term goals.
- So players can't grief or exploit them, or each other.
- So that content are rewards in and of themselves.

In other words, participants should be able to explore and impact a "world," or shared social space that features both content and interactive opportunities. They should be able to create and develop a unique identity within that world. They should see the bigger picture when it comes to doing work in the world—both an opportunity to escalate challenge and to continue working over time toward bigger results. The game must be carefully designed so that the only way to be rewarded is to participate in good faith, because in any game players will do anything they get the most rewarded for doing. And the emphasis must be on making the content and experience intrinsically rewarding, rather than on providing compensation for doing something that would otherwise feel boring, trivial, or pointless.

Do these principles work as effectively for real-world problem solving as for virtual-world problem solving? Absolutely. They are clearly the shared secret of the success for projects like Investigate Your MP's Expenses, Wikipedia, Free Rice, and Fold It!. In each case, the experience of participation is rewarding on its own merits, immersing a player inside an interactive world that motivates and rewards his or her best effort.

Gamers who have grown up being intensely engaged by well-designed virtual environments are hungry for better forms of engagement in their real lives.

They're seeking out ways to be blissfully productive while cooperating toward extreme-scale goals. They are a natural source of participation bandwidth for the kinds of citizen journalism, collective intelligence, humanitarian, and citizen science projects that we will increasingly seek to undertake.

As the examples in this chapter demonstrate, crowdsourcing games have an important role to play in how we achieve our democratic, scientific, and humanitarian goals over the next decade and beyond.

And more and more, these crowdsourcing games won't be just about online work or computational tasks. Increasingly, they will take us out into physical environments and face-to-face social spaces. These new games will challenge crowds to *mobilize* for real-world social missions—and they may make it possible for gamers to change, or even save, real people's lives as easily as they save virtual lives today.

CHAPTER TWELVE

Missions Impossible

Epic win /ˈɛpɪk ˌwɪn/ —

noun

1. *an unexpected victory from an underdog*
2. *something fantastic that has worked out unbelievably well*
3. *the greatest possible way for man to succeed at anything*

interjection

4. *an expression of happiness and/or awe at a highly favorable (and often improbable) event that has taken place: "Alright! Epic win!"*

—*from the Urban Dictionary*[1]

What the world needs now are more **epic wins**: opportunities for ordinary people to do extraordinary things—like change or save someone's life—every day.

"Epic win" is a gamer term. It's used to describe a big, and usually surprising, success: a come-from-behind victory, an unorthodox strategy that works out spectacularly well, a team effort that goes much better than planned, a heroic effort from the most unlikely player.

The label "epic" makes these kinds of wins sound rare or exceptional. But in the gamer world, they're not. Discussion forums are full of gamers sharing their most surprising and rewarding fiero moments. And they come in many different forms.

Some epic wins are about discovering we have abilities we didn't know we had. One action-adventure gamer writes: "After over an hour of attempting the ridiculously impossible office battle scene in *Indigo Prophecy*, which I was sure I'd never finish, I finally passed it, exhausted and wracked with awe. I did *that*? Epic win."[2]

Some are about upsetting other people's expectations of what's possible. A fantasy-football gamer writes: "I won the Champions League in *Championship Manager* coaching huge underdogs Malaga through a simulated season. Now that's epic. It's the most unlikely win ever."

And still others are about inventing new positive outcomes we hadn't even imagined before. A *Grand Theft Auto* player reports: "My epic win in GTA IV: Mountain biking to the top of the highest mountain from the city. Takes me 25 minutes real time. Just in time to see the sunrise."

What do these three different kinds of epic wins have in common? They all help us revise our notion of what constitutes a realistic best-case-scenario outcome. Whatever we thought the best possible result could reasonably be before, after an epic win we've set a new precedent: We can do more. It can get better.

With each epic win, our possibility space expands—dramatically. That's why epic wins are so crucial to creating sustainable economies of engagement. They make us curious about what more we can do—and as a result, we are more likely to take positive action again in the future. Epic wins help turn a one-off effort into passionate long-term participation.

Epic wins abound in gamer circles, for two reasons. First, in the face of ridiculous challenges, long odds, or great uncertainty, gamers cultivate extreme optimism. They have perfect confidence that even if success isn't *probable*, it's at least possible. So gamers' efforts to achieve an epic win never feel pointless or hopeless. Second, gamers aren't afraid to fail. Failing in a good game is at the very least fun and interesting; it can also be instructive and even empowering.

Extreme optimism and fun failure mean that gamers are more likely to put themselves in situations where epic wins can happen—situations where we take up unlikely missions and surprise ourselves with new awe-inspiring positive outcomes.

Ideally, the real world would present us with the same kind of intensely gratifying, save-the-world work flow we get from good games. But in real life, epic wins can be few and far between. We just don't have the same kind of carefully designed opportunities to surprise ourselves with our own super-powers.

We don't have an endless stream of opportunities to *do something that matters right now*, presented with *clear instructions*, and finely tuned to our *moment-by-moment capabilities*. Without that kind of creative and logistical support, there's no easy way to go after epic goals and successfully achieve them in our everyday lives.

Fortunately, a new genre of games called **social participation games** is trying to change that. They're designed to give players real-world volunteer tasks that feel as *heroic*, as *satisfying*, and—most importantly—as *readily achievable* as MMORPG quests. And as a result, a growing number of gamers are getting their hands dirty doing real-world good—and improving and saving real lives.

Take my friend Tom. He's a young math teacher who lives in Portland, Oregon. He usually gets his epic wins playing *Rock Band*, or, as he tells me, "any game where you get to play as Spider-Man or Batman." But recently, he started playing a social participation game called The Extraordinaries—and it has dramatically expanded his sense of his own potential.

THE EXTRAORDINARIES

The Extraordinaries is a Web and mobile phone application designed to help you do good in your spare time, wherever you happen to be. Created by a team of San Francisco–based designers, entrepreneurs, and activists, its primary objective is to make being heroic in the real world as easy as being heroic in a virtual world.

The game's motto is "Got two minutes? Be extraordinary!" Players can log in to the game from wherever they are and browse a list of "microvolunteer" missions that they can start and finish in literally just a few minutes. Each mission helps a real nonprofit organization accomplish one of its goals.

By design, The Extraordinaries' mission dashboard works almost exactly like the *World of Warcraft* log of available quests. You flip through available opportunities, and every mission you see comes with a story about why it will help save the world, along with a step-by-step explanation of how to get it done. There's never a shortage of important work to be done, and everything is designed to be doable by anyone willing to make a good-faith effort.

The first time my friend Tom logged in to The Extraordinaries, he immediately found a heroic mission he felt confident he could actually do. The mission was to use his iPhone to snap a photograph of a special "secret object," tag the photo with his current GPS location, and upload it to a database.

The secret object was a defibrillator, or AED—the device used to deliver a lifesaving shock to thousands of heart attack victims each year. The mission was designed by First Aid Corps, which is creating a map of every publicly accessible defibrillator in the world. As the organization explains in its mission's instructions:

> Each year, more than 200,000 Americans go into cardiac arrest—and within five minutes, the brain dies. Unfortunately, ambulances just can't always get there in time. Only those in the nearby vicinity can respond within that time.
>
> Government buildings, airports, schools, and more are installing defibrillators (shock pads) so that ordinary citizens can save lives in the event of an emergency. First Aid Corps is building a map of these devices with The Extraordinaries so that 911 can give someone a location to run to in the event of an emergency.

In other words, if you can find a defibrillator that isn't on the map yet, and if you successfully photograph and report it, you can help First Aid Corps save lives.

With good mission design—a focused task, a clearly defined context for action, a real window of opportunity—something previously impossible to achieve, like saving a life, becomes possible. That's the power of making volunteer work more like a game: players can be empowered to do amazing things, if their volunteer work is designed like a good quest.

In the First Aid Corps mission, the task of saving a life is presented just like a *World of Warcraft* quest. The instructions are straightforward, the reason for the mission compelling, and the task well within your ability level. If there's a defibrillator somewhere you plan to be today, then you can be a superhero right away. If not, you now have a secret mission everywhere you go, until you find the brokenhearted logo that is the international symbol for a defibrillator.

The defibrillator that Tom found was in an elevator bank at Portland State University, where he is completing a graduate degree in math education. "I've looked past it while waiting for the elevator for years," he told me afterward. "Suddenly it was relevant, and I was glad to have this random secret info." Of course, it wasn't secret information at all; the defibrillator was in plain public view. But Tom's words here reveal just how effective The Extraordinaries' promise really is: to give you a real chance to feel like a superhero, on a secret mission to save the world.

After Tom completed his mission, the win was scored on The Extraordinaries' activity board for every other player to see: "Tom H mapped a defibrillator and helped to save lives."

Later, Tom e-mailed me the news. "It was like a lifesaving scavenger hunt," he told me. "Inherently awesome. Massive epic win." The defibrillator mission was an epic win because, until that morning, Tom had no idea he had knowledge that could help save a life. He had a secret power he didn't know about—and he was given a real opportunity to put that power to use.

What happens next? If Tom's defibrillator does help save a life, he'll know. The First Aid Corps updates its global map with links to news stories about each defibrillator's usage. If "Live Saved" pops up next to *your* AED location, then you know that the AED you found really *has* helped save the day. Right now, it's up to players to proactively check the status of their AEDs. But it's easy to imagine a platform like The Extraordinaries evolving to push updates directly to players via text message or social network update whenever their small act of good helps accomplish something bigger. In that case, the small yet epic win of discovering and sharing a defibrillator's location could lead to an even bigger epic win down the road.

The call to action of The Extraordinaries—"Be extraordinary!"—is really just another way of saying: Surprise yourself with how much good you can do. Redefine what your best possible outcome for the day could be. It's not that we don't have the ability to do good for others. It's just that no one has shown us how fast, easy, and addictive it can be to tackle what feel like missions impossible.

By the fall of 2009, within just a few months of its launch, The Extraordinaries had become a small but growing social network, with more than thirty-three hundred members who had collectively completed more than twenty-two thousand missions on behalf of more than twenty nonprofit organizations. That's an average of seven epic wins per member. Judging from just that statistic alone, the app clearly isn't the most addictive experience in the world yet. But it's doing extremely important work: it's showcasing the potential for more epic wins, every day, for everyone.

Which brings us to our next fix for reality:

 FIX #12: MORE EPIC WINS

> Compared with games, reality is unambitious. Games help us define awe-inspiring goals and tackle seemingly impossible social missions together.

Why do we need more epic wins in our everyday lives? Right now, as a planet, we are collectively facing some of the most incredible odds in our history: climate change, global economic crises, food insecurity, geopolitical instability, and rising rates of depression worldwide. Emphatically, these are problems that cannot be solved online. They require real-world action, not just online interaction.

The exciting promise of a project like The Extraordinaries is that we can do more than pick the brains of gamers.[3] We can harness the social participation of the masses.

Social participation means using more than our minds: it requires throwing our hearts and our bodies into action. So the challenge that lies ahead is to design **social participation tasks** (SPTs) to stand alongside the growing number of human intelligence tasks (HITs) that currently make up the majority of online crowdsourcing projects: transcribing and subtitling videos on DotSUB, for example, analyzing an MP's receipts in Investigate Your MP's Expenses, or even simply evaluating an idea for a new product name as "good" or "bad." What these efforts all have in common is that they appeal primarily to our cognitive, rather than our emotional and social, capabilities.

HITs are, without a doubt, important work, but we are more than just thinking machines. We are human beings capable of reaching out to others, feeling empathy, recognizing need, showing up, and making a difference in someone else's life. We have *social powers*, and we can mobilize them for good—in real-world spaces, not just online spaces. All we need is the right kind of mission support.

Consider one more mission from The Extraordinaries game—it's my personal favorite, the one that made me feel the biggest epic win. This one is a social participation task for Christel House, an organization dedicated to helping children living in poverty get the education, nutrition, health care, and mentorship they need to become self-sufficient, contributing members of society. And it's a perfect example of a mission that takes advantage of some of our key social powers: the ability to empathize, advise, and provide positive emotional support.

The mission is simple: Write a short text message of good luck to a child about to take a potentially life-changing standardized test. You can choose whether to send your message to a child in the United States, Mexico, Venezuela, South Africa, or India. Christel House will ensure your online message gets into the hands of a real student, in a physical classroom, moments before he or she takes the test. Nathan Hand, the development associate for Christel House who helped design this social participation task for The Extraordinaries, explains it this way:

> All these kids around the world have at some point, in every country, some sort of standardized test that they need to pass. Sometimes it makes or breaks graduation, sometimes it makes or breaks them getting into the next grade level — it depends on the country — but no matter where the child is, it's a lot of pressure, and they spend their whole life preparing for it. What we're trying to do is basically crowdsource the pat on the back.[4]

I chose to write my good-luck message to a student in India. I shared my favorite trick: "Before you start the test, smile as wide as you can! If you get stuck on a hard question, stop, and smile!" I knew from scientific research that smiling even when you don't feel like it can actually trigger real feelings of confidence and optimism.[5]

As I clicked send, I pictured a young student in India taking my advice. In that moment, I felt meaningfully connected to another human being I had hardly any hope of meeting or speaking to otherwise. I had real hope that I was able to reach out to another person in a time of difficulty and give them support that mattered. In other words, I had exactly the experience Hand describes as the goal of the Christel House Extraordinaries mission: "People literally, in a matter of seconds, can have a meaningful engagement with a kid in need through us. They have the warm glow, then they remember us, and they remember those kids, and that's what it's about."[6]

Before I found this mission, I'd had no intention of trying to help a child halfway around the world ace an important, potentially life-changing test. It's not just that it wasn't on my to-do list. It wasn't on my *possible to-do* list. The good game design of the Christel House mission changed that: it made it incredibly easy to play a helpful role in a stranger's life. It showed me a capacity to help I didn't know I had. It gave me goose bumps.

That's an epic win already, because it changes our perspective of who we are, how much we care, and what we're capable of doing for others.

SOCIAL PARTICIPATION GAMES are innovating human potential. They are augmenting and expanding our capabilities to do good—and revealing our power to help each other, in the moment, wherever we are.

The Extraordinaries is a perfect example of how epic wins can be integrated into our everyday lives, and how we can generate more participation bandwidth worldwide. But it's not the only example—far from it. Let's take a look at two more extraordinarily ambitious projects that are attempting to harness the social capacity of crowds: Groundcrew, a mobile people-organizing platform that allows you to make real-life wishes come true, and Lost Joules, an online energy conservation game that invites you to make virtual currency wagers on just how much social good other players can accomplish.

GROUNDCREW—POWERING THE MOBILE
COLLABORATION ECONOMY

The best way to explain the wish-granting Groundcrew project, developed by Cambridge, Massachusetts–based social entrepreneur Joe Edelman, is by looking first at the project's inspiration, one of the best-selling computer games of all time: the life-simulation game *The Sims*.

When you play *The Sims*, your job is to keep your simulated people healthy and happy. You keep them healthy by tending to their physical needs: feeding them, putting them to sleep, making sure they shower and use the bathroom regularly. You keep them happy by fulfilling "wishes."

As *The Sims 3* guide explains, "Sims come up with small wishes each day that they would love for you to help them fulfill. Fulfilled wishes boost your Sim's mood and award Lifetime Happiness points."[7] Sims express their wishes to you via the "wish panel," a kind of head's-up display that shows you exactly where each Sim character is, what they want, and how you can get it for them. For example: "The night sky is beautiful and mysterious. Your Sim wants to explore the logical patterns of the stars. Use a telescope: Worth +150 happiness points." The wish panel gave Edelman his breakthrough idea. "Real people have wishes just like the Sims," Edelman says. "The problem is, we just don't know what those wishes are, or how we can help. What if we could receive real-time alerts about how to make real people happy?"[8]

So Edelman started a software company called Citizen Logistics to build a wish panel for real people. His vision: to make it as easy to satisfy the everyday wishes of other human beings as it is to improve the lifetime happiness score of our favorite Sims.

The logistics part would be key, Edelman realized. Right now, it's not easy to find out what we can do, in the moment and where we happen to be, to make someone else's day. So he set out to create a new system that would map real-time wishes onto our local environment. The concept has three key features.

First, a player should be able to log in to the system and see everyone within walking, public transportation, or driving distance who has a wish at that very moment. Meanwhile, players with wishes of their own should be able to see a map of all available "agents" in the area who are up for a quick wish adventure, and they should be able to push their wish at available players directly, via text messaging. Finally, the first player to successfully fulfill someone else's wish in time should earn reputation points to indicate that they are a trusted wish fulfiller. This would allow them access to fulfill more challenging wishes over time. Even better, they could later spend their earned reputation points to mobilize and reward other players for fulfilling their own wishes.

Edelman wasn't sure the idea would work, but he believed it was worth trying—and so he built a test platform and invited friends and colleagues in the Cambridge area to use it for anything they wanted. To his delight, the idea did work, and right away. On the first day the system was live, what Edelman calls Wish #1 was granted. As he tells the story:

A woman was at a dance rehearsal in a basement somewhere in Boston. She was completely exhausted, she couldn't leave rehearsal, and she was dying for a latte so she could keep dancing. That's the wish she posted on Groundcrew: "Help. I need a latte."

At this exact moment, someone else in Boston is watching the system. He sees her wish. And he realizes he's only a few blocks away from the dancer. It feels like fate. He can do this! He knows how to order a latte! He can save the day!

Five minutes later, he walks into the basement and declares, "I have a latte!" as if it were the most important thing in the world. And it is the most important thing in the world, at that minute, to that dancer! She is overjoyed. She says it's the best latte she's ever had. He feels like a superhero. All of this all transpires within a few minutes of the wish alert.[9]

Okay, so getting someone a latte isn't exactly the most world-changing effort you could make.

Or is it?

Edelman likes to tell the latte story, even if it seems like a trivial wish, because for him it perfectly represents the new kind of epic win that is possible in a world where more and more people are willing to use their mobile phones to broadcast where they are and what they need. The win is nothing less than an augmented human capacity to do good and feel good every single day, by making better use of each other's spare time. As Edelman puts it, "We can love a lot more people when we can make their wishes come true in seconds. . . . We can love people when we know what they need."[10]

This isn't just a warm, fuzzy fantasy. Edelman is talking about reinventing our idea of everyday economic systems of give-and-take. "The normal way of getting a latte is a cold, economic exchange," he says. "We walk into a café alone every day, and we give up our hard-earned cash to get it. But this latte was different. This latte was love. This is about inventing a different way, a better way, of getting what we need, every day."[11]

⟳ YOUR GROUNDCREW MISSION

"What if real life were more like a game? In recent years, virtual community has gotten easier than physical community. Computer games provide expertly designed entertainment and pleasure . . . but when we have to deal with our real lives, we're all alone. When will participating in the real world and dealing with real issues be just as adventurous, easy, collaborative, and fun?

Our definition of community is actual people, in vicinity to each other, thinking about each other's needs and helping each other, in person and on the ground. We want to see a decrease in loneliness, helplessness, isolation, and needless expense across America and across the world. We want an increase of enjoyment, adventure, conviviality, sharing, and mutual support.

We seek to assist the human desires to be available for one another, to be good to one another, to rejoice in one another, to make good use of our ecological and social resources, and to engage with life in ways that are real, deep, and unpredictable."[12]

—Groundcrew founder Joe Edelman

Imagine for a moment what kinds of needs you might express in the course of your everyday life: I'm bored. I'm lost. I'm hungry. I'm lonely. I'm nervous. What could you wish for to fill those daily needs?

I can think of lots of small wishes I would make:

I'm stressed at work and want to play fetch with a dog to help me relax. Please walk your dog here, now!

I'm flying out of SFO in the morning and I want to read your old copy of the new Dan Brown book. If you are going to be in Terminal 1 between seven and eight a.m., please bring it to me!

I'm giving a public talk at the university tomorrow; please come and try to spark a standing ovation at the end of it, because my parents will be in the audience and I want them to be proud.

None of these wishes would change my life. But they would completely change my notion of how to get what I want from life—and more importantly, how to share what I have with others.

Indeed, Groundcrew represents the potential for an entirely new kind of economy, built around the exchange of three intrinsic rewards: the happiness that comes from doing good, the thrill that comes from accomplishing a challenging mission, and the satisfaction of accumulating points that signify something real and wonderful—your ability to make other people's wishes come true, and your future chances at having your own wishes fulfilled.

There's no inherent limit to this new engagement economy: all three of these rewards are infinitely renewable resources. And Groundcrew's original virtual currency, the PosX (short for "positive experience") reputation system, makes it possible, for the first time, to accumulate, quantify, and exchange these intrinsic rewards.

"The availability of cheap, networked, programmable devices is as big a deal for human economics as the invention of paper money and coins were," Edelman explains. "It gives us, for the first time, the opportunity to *change the rules of the game,* to *tune the incentives,* and to create much more flexible *access to resources*—including other people—all without creating the huge bureaucracies and informational inefficiencies associated with previous attempts.

"While we continue to argue about capitalism and socialism, for the first time a third option is really possible. Right now, we have an opportunity to make things more equitable, more sustainable, more intimate, and also more beautiful and fun. When incentives match up better with our deep human desires, life becomes more enjoyable, adventurous, and fulfilling."[13]

The more missions you participate in, the more PosX you receive. But that's just the start of the economy. Someone who enjoys completing your mission can also give *you* PosX for giving them the chance to do something that feels good. It's an incredibly smart, radical idea, and it's derived directly from the economic model of the game industry. People are happy to pay money— buying and subscribing to games—for the opportunity to do hard work that is intensely rewarding. And a truly sustainable economy of real-world engagement should strive to harness this market for better, more rewarding work.

In Groundcrew, you can be paid in virtual currency for doing good work—but you can also pay others in that same virtual currency for giving you good work to do. This will create a market for satisfying social participation tasks. It will mean many more people trying to design real-world missions that you can achieve right away, giving you real hope of success, increasing your social connectivity, and giving meaning to an otherwise boring day. This is pivotal: we can't have more epic wins in daily life unless smart people are contributing good SPTs to our collective save-the-world work flow.

Of course, some wishes are more urgent than others. Groundcrew is currently working with AARP, the nonprofit organization for Americans over the age of fifty, to empower agents to "make a difference in the lives of elders near you." Groundcrew agents receive SMS and e-mail alerts with special elder-focused SPTs: help with transportation, grocery shopping, light housekeeping, or just companionship. Because these missions involve intimate interaction with a potentially vulnerable population, not just any agent can undertake these missions—only the most trusted agents (who have racked up enough PosX and also submit to criminal background checks) are eligible for these "high-trust" SPTs.

I have to admit—I'm partial to the kinds of intimate exchanges first imagined by Edelman when he invented his real-world wish panel. I think that improving each other's daily lives by making small, one-to-one efforts in our spare time could dramatically improve global quality of life and make more sustainable, efficient use of our material resources. But Groundcrew is also a scalable project, capable of harnessing huge crowds for a single wish.

Indeed, since Edelman started developing the Groundcrew platform, he has evolved his vision, so that players can help fulfill not only individual wishes, but also organizational goals. Like The Extraordinaries, Groundcrew has started to partner with existing institutions to find volunteers for a variety of nonprofit, activist, and political organizing efforts.

Groundcrew's first crowd mobilization efforts was for Youth Venture, an organization that encourages young people to take social action and start "businesses for good" in their local communities. One signature Youth Venture initiative, Garden Angels, coordinates efforts to create community gardens, with the goal of distributing the fruits and vegetables grown in them to people who need it most. Many people know about and support this effort, but they don't have an easy way to help. That's where Groundcrew comes in—to create an epic win work flow for community gardens.

Garden Angels is using Groundcrew to organize large-crowd events in local gardens, like soil turning and gleaning, as well as to find volunteers for small, everyday activities: weeding, watering, and checking on the security of the gardens. Instead of planning an event and hoping volunteers show up, Youth Venture can wait for a critical mass of Groundcrew players to signal their availability for a mission, then throw an impromptu event strategically timed to harness as much participation as possible. Meanwhile, players don't have to schedule their volunteer efforts in advance; they can sign up to receive text messages when a small task needs to get done in a garden that happens to be nearby. (Whenever players check in to report their location, the system searches for nearby tasks.) In testing Groundcrew with projects like Garden Angels, Edelman reports, they've already seen on average "a hundred times increase in the availability of volunteers for projects."[14]

That's how to increase our collective social participation bandwidth: by empowering one hundred times as many people to make heroic efforts in their spare time.

WHETHER WE'RE HELPING individuals or helping big organizations, our notion of how much social engagement we can expect from an ordinary person increases dramatically when gameful thinking meets smart technology. That's why many social participation games are taking advantage not just of good game design, but also of leading-edge technologies that make it easier to plug individual action into epic contexts. So far, mobile phones have been at the forefront of this effort—but they're not the only way to add epic wins to our daily lives.

Consider Lost Joules, which promises to be the world's greenest computer game. It helps us get epic wins in our own homes by turning our electricity meters into game controllers.

LOST JOULES

Imagine it's Friday afternoon, and I have an important favor to ask of you. For the good of the planet, you need to try to conserve as much energy as possible at home this weekend. Turn off your lights earlier, charge your electronics less frequently, unplug your toaster, hang your clothes on a clothesline instead of using the dryer. How hard would you try to do me that favor?

Now let's say I told you that I had a hundred dollars riding on your ability to reduce your overall energy usage by at least 20 percent this weekend. How hard would you try to help me win?

Finally, one more scenario. This time, I've got a hundred dollars riding *against* your ability to reduce your overall energy usage this weekend by 20 percent. How hard would you try to prove me wrong?

I'm not able to bet on your energy usage yet—but when the Lost Joules game launches, I will be. It's an online stock market game that lets players make wagers (in virtual currency) on each other's real-world energy usage. Players have a strong motivation to place good bets: if they win the bet, they'll be able to spend the virtual currency they earn inside the Lost Joules "virtual theme park," which will house a number of FarmVille-type games. The more energy bets you win, the more powerful and rich your Lost Joules avatar will become.

The game works with smart meters, home electricity meters that are connected to the Internet. Smart meters allow you to monitor and analyze how and where your energy is being consumed—they can even calculate exactly how much each appliance in your house is costing you. Studies have shown that having this kind of feedback makes it much easier to reduce energy consumption: on average, a smart meter user will be able to decrease his or her consumption permanently by 10 percent.[15] And that's without friends, family, and strangers cheering you on, or trying to beat your best effort. Can you imagine how much more energy could be saved if using smart meters was turned into a good game?

Lost Joules is set to find out. The application collects personal smart-meter data from players and challenges them to achieve concrete, energy-saving missions. Then it makes that data public to other players—who will place bets on your ability to achieve energy-saving missions. If they think you can do it, they'll bet with you—and if they doubt you, they'll invest in someone else. The players who achieve the most missions regularly will become superstars in the Lost Joules world, generating returns not only for themselves, but also for everyone who cheers them on.

By creating a sense of urgency, presenting a clear challenge, and adding a layer of social competition, the game turns what would otherwise feel like an ordinary, mundane effort to do a bit of good into an extraordinary effort. Suddenly, turning off an appliance becomes an epic win, with multiple rewards: emotional rewards, like more fiero and better social connectivity, *and* virtual rewards, in the form of game-world currency.

It's a very big, very new idea. Lost Joules is seeking to create a sustainable engagement economy around what is currently an unsustainable energy economy. To motivate people to consume less nonrenewable energy, it offers them the opportunity to consume completely renewable emotional and virtual rewards.

It's also creating a new way of helping to save the world: by investing our social attention in people who are doing good. As the game's cocreator Richard Dorsey likes to say, "Wouldn't it be cool if every time we unplugged an appliance or flipped a switch, somebody noticed?"[16] By turning energy saving into a massively multiplayer experience, Lost Joules takes advantage of the network effect: it amplifies my private epic wins into spectacular social achievements.

Of course, many people won't want their energy consumption to be scrutinized and wagered on by the playing public. But given the history of increasing public disclosure on the Internet—from blogs to videos to social network to real-time status updates—it's a safe bet that the lure of being lauded in the public spotlight will attract plenty of players. And thanks to the game's two-tier design, even people not ready to expose their own energy consumption can help drive energy-saving behavior just by making a virtual investment.

In this way, Lost Joules represents an important design innovation in the social participation game space. It's creating two different kinds of equally important social participation tasks, for people with smart meters and people without smart meters.

First, and most obviously, players with smart meters can tackle the social participation task of reducing their energy consumption. This is the core "do-good" mission of the game. But there's also the SPT of lavishing our attention on each other's good acts. People who don't have access to smart meters yet can still play the game, by making wagers on players who do have smart meters. And this is a real contribution to the common good, since it creates social rewards for the energy savers. Everyone likes to feel valued; Lost Joules uses virtual currency to help us show just how much we value the world-changing contributions of others.

So what's the best-case-scenario outcome for a game like Lost Joules? Games are a major driver of technology adoption; people are often more willing to try new technologies when there's a good game attached. And getting people to try smart-meter technology is increasingly important, as we try to become more informed, efficient consumers of energy. Smart meters have been proven remarkably effective at changing our energy consumption behaviors for the better. The more people who use them, the better.

In the bigger picture, the real potential of Lost Joules is to demonstrate how to make better use of the abundant emotional and virtual rewards that games provide to motivate change-the-world behavior. Right now, it's easier and more fun to be a superhero in a video game than it is to help solve real global problems in everyday life. But social participation games like Lost Joules are starting to tip the balance: soon, we may find ourselves able to do both at the same time.

The three projects described in this chapter — The Extraordinaries, Ground-crew, and Lost Joules — are just starting to unfold. They are all highly speculative, still in development, with modest if any results so far. They are beyond leading edge. They're *bleeding edge*: so new, there's significant risk that they will fail.

In fact, there's a very good chance some of them may even wind up being examples of an epic fail rather than an epic win. But, as any good gamer knows, failure can be both rewarding and empowering, if you learn from your mistakes. Testing our potential to do more than we thought possible brings us closer to achieving it someday. As the familiar saying goes, "Even if you fall flat on your face, you're still moving forward."

Epic wins, when connected to real-world causes, help us discover an ability to contribute to the common good that we didn't know we had. They help us upset other people's expectations of what is possible for ordinary people to accomplish in their spare time. And they help us set goals that would have seemed ludicrous—impossible—before we had so many volunteers so well equipped to help each other, and so effectively mobilized.

In short, social participation games are turning us into superheroes in our real lives.

And every superhero needs superpowers.

What kind of superpowers do we need most? Collaboration superpowers— the kind that enable us to combine forces, amplify each other's strengths, and tackle problems at a planetary scale.

CHAPTER THIRTEEN

Collaboration Superpowers

By the age of twenty-one, the average young American has spent somewhere between two and three thousand hours reading books—and more than *ten thousand* hours playing computer and video games.[1] With each year after 1980 you're born, these statistics are increasingly likely to be true.

To put that number in perspective, ten thousand hours is almost exactly the same amount of time an average American student spends in the classroom from the moment they start fifth grade all the way through high school graduation—if they have perfect attendance. In other words, as much time as they spend learning reading, writing, math, science, history, government, geography, foreign languages, art, physical education, and so on over the course of their middle school and high school careers they spend teaching themselves (and each other) to play computer and video games. And unlike their formal education, which diffuses their attention across myriad different subjects and skills, every single gaming hour is concentrated on improving at just one thing: becoming a better gamer.

With ten thousand hours under their belts by age twenty-one, most of these young people will be more than just good gamers. They'll be *exceptionally* good gamers.

That's because ten thousand hours of practice before the age of twenty-one, according to at least one theory, is the number one predictor of extraordinary success later in life.

Malcolm Gladwell first proposed the ten-thousand-hour theory in his best-selling book *Outliers: The Story of Success*. In *Outliers*, Gladwell reports on the life stories of high-achieving individuals, from violin virtuosos to all-star hockey players to Bill Gates, and he finds that they all have one autobiographical fact in common. By the age of twenty, the top performers in any given field had each accumulated at least ten thousand hours of practice at the one thing that eventually made them superstars. Meanwhile, the runners-up—the second tier of successful, but not extraordinarily successful, musicians, athletes, technologists, businesspeople, and so on—had on average eight thousand or fewer practice hours each.

Natural talent matters, of course, but not as much as practice and preparation. And, according to Gladwell, ten thousand hours of practice and preparation appears to be the crucial threshold, marking the difference between simply being good at something and becoming extraordinary at it.

This means that we are well on our way to creating an *entire generation* of virtuoso gamers. Every young person who achieves ten thousand hours of gaming practice will be capable of extraordinary success in gaming environments later in life.

It's potentially an unprecedented human resource: hundreds of millions of people worldwide who are going to be exceptionally good at the same thing—whatever it is games make us good at.

Which brings us to the million-dollar question for the future: What, exactly, are gamers getting good at?

I've been researching that question for nearly a decade, first as a PhD student at the University of California at Berkeley and later as the director of game research and development at the Institute for the Future. Over the years, it has become increasingly clear to me that gamers—especially online gamers—are exceptionally skilled at one important thing: collaboration. In fact, I believe online gamers are among the most collaborative people on earth.

Collaboration is a special way of working together. It requires three distinct kinds of concerted effort: *cooperating* (acting purposefully toward a common goal), *coordinating* (synchronizing efforts and sharing resources), and *cocreating* (producing a novel outcome together). This third element, cocreation, is what sets collaboration apart from other collective efforts: it is a fundamentally *generative* act. Collaboration isn't just about achieving a goal or joining forces; it's about creating something together that it would be impossible to create alone.

You can collaborate to create just about anything: a group experience, a knowledge resource, a work of art. Increasingly, gamers are collaborating to create all of these outcomes. In fact, they're collaborating even when they're competing against each other to win. More and more, gamers are collaborating even when they're playing alone.

It seems counterintuitive: how can you collaborate with someone when you're actively opposing them? Or even harder to imagine: how on earth can you collaborate all by yourself? But in fact, online gamers are increasingly doing both, thanks to two factors: the fundamentally collaborative aspects of playing any good game, and new game technologies and design patterns that support entirely new ways of working together.

The Evolution of Games as a Collaboration Platform

Since ancient times, gaming with others has always required making a concerted effort to collaborate. This is true of dice games, card games, chess, sports, and any other kind of multiplayer game you can think of.

Every multiplayer game begins with a cooperative agreement. Gamers agree to play by the same rules and to value the same goal. This establishes a **common ground** for working together.

Games also require us to coordinate attention and participation resources. Gamers must show up at the same time, in the same mind-set, to play together. They actively focus their attention on the game, and they agree to ignore everything else for as long as they're playing. They practice **shared concentration** and **synchronized engagement.**

Gamers depend on each other to play as hard as they can, because it's no fun winning without a challenge. In this way, gamers foster **mutual regard.** Out of respect for each other, they put in their best effort, and they fully expect to encounter a worthy partner or adversary.

Gamers rely on each other at all times to keep the game going, even if it's not working out in their favor. Whenever they see a game through to completion, gamers are honing their ability to honor a **collective commitment.**

Perhaps most importantly, gamers actively work together to make believe that the game truly matters. They conspire to give the game real meaning, to help each other get emotionally caught up in the act of playing, and to reap the positive rewards of playing a good game. Whether they win or lose, they're creating **reciprocal rewards.**

In short, good games don't just happen. Gamers work to make them happen. Any time you play a game with someone else, unless you're just trying to spoil the experience, you are actively engaged in highly coordinated, *prosocial* behavior. No one forces gamers to play by the rules, to concentrate deeply, to try their best, to stay in the game, or to act as if they care about the outcome. They do it voluntarily, for the mutual benefit of everyone playing, because it makes a better game.

This is true even in games that involve fierce competition. Consider the origins of the English word "compete": it comes from the Latin verb *competere*, which means "to come together, to strive together" (from *com-*, or "with," and *-petere*, meaning "to strive, seek"). To compete *against* someone still requires coming together *with them*: to strive toward the same goal, to push each other to do better, and to participate wholeheartedly in seeing the competition through to completion.

That's why today competitive online gamers—even after they've been virtually beaten, bloodied, or blasted by each other—thank each other afterward by typing or saying "GG," short for "good game." It's a grateful acknowledgment that, regardless of who wins or loses, everyone in a good game has tried hard, played fair, and worked together. That's the fundamental act of collaboration at the heart of every good multiplayer game: the active and concerted creation of a positive experience. Gamers don't just play a good game. They *make* a good game.

In fact, the ability to make a good game together has recently been identified by researchers as a distinctive human capability—indeed, perhaps *the* distinctive human capability. The developmental psychologist Michael Tomasello, author of *Why We Cooperate* and codirector of the Max Planck Institute for Evolutionary Anthropology in Leipzig, Germany, has spent his career devising experiments to investigate what kinds of behaviors and skills set humans apart from other species. His research suggests that the ability to play complex games together, and to help others learn the rules of a game, represents the essence of what makes us human—something he calls "shared intentionality."[2]

Shared intentionality, according to Tomasello, is defined as "the ability to participate with others in collaborative activities with shared goals and intentions."[3] When we have shared intentionality, we actively identify as part of a group, we deliberately and explicitly agree on a goal, and we can understand what others expect us to do in order to work toward the goal. Tomasello's research reveals that, in comparison with humans, other intelligent social species like chimpanzees simply do not appear to have shared intentionality. They don't have the natural instinct and ability to focus their attention on the same object, coordinate group activity, assess and reinforce each other's commitment to the activity, and work toward a common goal.

Without the distinctly human capacity for shared intentionality, we couldn't collaborate; we would have no idea how to build common ground, set group goals, or take collective action. According to Tomasello, children are capable of shared intentionality at a very early age. His evidence: their natural ability to play a game with others, and their ability to recognize when someone isn't playing the game in a way that favors the group.

In one of Tomasello's key experiments at the Max Planck Institute, children between the ages of two and three are taught to play a new game together — either a dice game for the two-year-olds or a building-block game for the three-year-olds.[4] Then a puppet controlled by another experimenter joins the game and plays it incorrectly, according to its own made-up rules. Tomasello and his colleagues report that children immediately and universally object to this bad game behavior and attempt to correct the puppet, in order to keep the game successfully going — even though they haven't been instructed to do so. This behavior was more "vociferous" among the three-year-olds, according to the published findings, but clearly widespread among the two-year-olds as well. We are able to make a good game together — and we are inclined to do so from nearly the moment we are born. We have a hardwired desire and capacity to cooperate and coordinate our actions with others, to effectively immerse ourselves in groups, and to actively cocreate positive shared experiences.

And yet this desire can be diminished and our natural abilities weakened or eventually lost, Michael Tomasello argues, if we grow up in a culture without sufficient opportunities to nurture and develop it.

If we are to achieve our human potential to be extraordinary collaborators, he urges, we must immerse ourselves in high-collaboration environments — and we must encourage young people to spend as much time as possible participating in groups that encourage and value cooperation. Fortunately, as online and multiplayer games become more and more central to global popular culture, we have all the encouragement we need to practice our natural collaboration abilities. Multiplayer and online games strengthen our capacity to build and exercise shared intentionality.

Every time we agree to play a game together, we are practicing one of the talents that makes us fundamentally human.

THIS IS NOT to suggest that online gaming today is one giant cooperative utopia. The kill-or-be-killed adrenaline rush of player vs. player environments can easily overshadow the very real undertones of cooperation and collaboration that otherwise exist. Graphic violent content, combined with the anonymity of the Internet, doesn't necessarily inspire camaraderie among strangers. That's why toxic social interactions can and do erupt in hard-core, or especially competitive, communities, as normally playful gamer behaviors like taunting and trash-talking get out of hand.

Even in friendlier matches, many gamers care very deeply about whether or not they win. They're seeking that fiero moment and wind up feeling disappointed or angry if they lose. In that case, even the fundamentally collaborative spirit of making a good game together can't completely alleviate the sting of loss.

Yet despite all these potentially mitigating factors, gamer culture is moving insistently in the direction of more shared intentionality, not less. For the past few years, cooperative, or *co-op*, play and *collaborative creation* systems have consistently remained the most celebrated trends in gaming.

In **co-op mode,** gamers work together to defeat an AI opponent and to increase each other's scores, rather than competing against each other. Classic examples of co-op play include *Rock Band* and the first-person shooter series *Left 4 Dead*. Although there are competitive elements to both games, the primary focus is on working together to achieve a goal.

Even in game series that have previously specialized in single-player and player vs. player experiences, co-op mode is becoming more and more central. The counterterrorist-themed *Call of Duty: Modern Warfare 2*, the fastest-selling entertainment product in world history—it grossed $550 million in five days, more than any book, movie, album, or other game ever produced—has been particularly praised for its new Spec Ops mode, a series of twenty-three extremely challenging missions designed to be played cooperatively with a friend.

The industry's increased attention to co-op mode represents an extremely significant development in gaming culture. It's a recognition that many gamers are happier tackling challenges together than taking on each other as opponents. Co-op games deliver all the emotional rewards of a good game, while helping gamers avoid activating the negative emotions that can come with highly competitive play: feelings of aggression, anger, disappointment, or humiliation. That's why it's not surprising that surveys and polls repeatedly have shown that, on average, three out of four gamers prefer co-op mode to competitive multiplayer.[5]

Game developers aren't just designing more co-op play; they're also creating new **real-time coordination tools** to help us find the right people at the right time to cooperate with. The Xbox Live platform, for example, enables players to monitor who else in their social network is logged in to the game console, what they're playing at the moment, and what other games they have in their library to play with you. You can browse your friends' records of game achievements and compare them against your own—which helps you figure out who would make a good partner on a given mission or in a particular game. You can also receive alerts on your mobile phone or your computer whenever, for example, a friend logs in to Xbox Live to play a game or he or she unlocks an achievement. As a result, Xbox gamers have an unusually high level of awareness of what potential coplayers are doing at any given time, what they're good at, and what resources they have to play with. The ambient awareness dramatically amplifies their ability to coordinate good gameplay.

Meanwhile, in **collaborative creation systems**, gamers get to create their own digital content, in order to build up their favorite worlds for the benefit of other players. Take *Little Big Planet* for example—it's one of the most acclaimed collaborative creation games released in recent years. In the traditional "play mode" of the game, you cooperate with up to three friends to traverse the game world and collect game objects together—stickers, gadgets, toys, and craft and building materials. At any time, you can switch from play mode to "create mode"; here you find yourself in a collaborative building environment called Popit, in which you can design your own original action-adventure landscapes out of the objects and materials you've already collected. It's a level-building system that you might call the game-design equivalent of Google Docs. Multiple people can view and edit the level at the same time; it can then be shared, or "published," to the rest of the world.

Within a year, more than 1.3 million player-created levels had been published by *LBP* players. Compare this epic number with the relatively small number of official *LBP* levels: forty-five. Collectively, the *LBP* player base has dramatically expanded the playable *LBP* universe by a factor of nearly thirty thousand. As one games journalist observed on the one-year anniversary of the game's release, "[It] would likely take multiple lifetimes to play through every single creation out there."[6]

The ability to create your own levels and share them with other players was the signature selling feature of *Little Big Planet*. But increasingly, successful game series are offering similar systems as a "value-add," in order to give players more explicit collaboration opportunities. For example, *Halo 3* introduced the new Forge system, which invites players to design their own original multiplayer *Halo* levels, or "maps," by customizing what weapons, vehicles, and tools are distributed where. Like LBP's Popit system, players can upload and share their custom configurations with each other, and using the Forge tools, it's possible to create literally billions and billions of different maps. So instead of being restricted to a finite number of play environments, the *Halo* community can keep the game going, increasing and diversifying the playing challenges for each other indefinitely.

It's not easy to design a good world, of course. So alongside the growing collection of collaborative creation systems, there are also a growing number of player-created guides to creating better levels and maps. Take, for example, the Forge Hub, a resource for becoming a better *Halo 3* world builder. It offers extensive tutorials in various mapmaking skills and curates player-created maps into different collections. It's a natural extension of the knowledge sharing and collective intelligence culture already taking place on the more than ten thousand player-created game wikis. Gamers aren't just making each other better players; they're making each other better designers.

But perhaps the most unusual innovation in gamer collaboration culture in recent years is the notion of the **massively single-player online game.** It's a twist on the traditional massively multiplayer online concept—and, on first impression, it sounds like an impossible paradox. How can you have a "massively" single-player experience when by definition a single-player experience occurs alone?

The inventor of the term is Will Wright, the famed creator of *SimCity* and *The Sims* games. He coined it to describe his 2008 game *Spore*, a simulation of the universe that invites players to design a galaxy from scratch, starting with a single-cell creature and evolving it up into a land-dwelling species, then into tribes, complex civilizations, and ultimately a space-faring, planet-designing megacivilization.

All *Spore* gameplay is single-player: an individual controls all the simulation details and conducts all the fighting, mating, crafting, and exploring alone. There are no other players in the simulated ecosystem; everything in the world is controlled by artificial intelligence. So what makes it *massively* single-player, as opposed to simply single-player? A very large percentage of the content in each player's game world — the other creatures you encounter and the civilizations you visit — has been created by other players who have contributed them to the online Sporepedia, a massive database of ecosystem content. When you play *Spore* online, your computer checks the Sporepedia for new and interesting content and downloads it into your personal Spore ecosystem, making your game world a mix of your own original contributions and those of many, many others.

Although there is no direct interaction with other players, you indirectly collaborate with each other to invent the *Spore* universe. You can randomly populate your world with other players' creations, or you can handpick creations you like from the Sporepedia. You can even subscribe to a Sporecast, which will automatically update your game world with new content created by your friends or favorite players.

Players use Sporepedia and the *Spore* forums and wikis to learn what other players are making and to improve their own creation techniques. They don't collaborate in the real-time gameplay, but ultimately the world that players help design is a collaborative product: a unique combination of each player's own creations mingled with content from hundreds, thousands, or even millions of other players, depending on how far they get in the game and how much content they choose to download.

A massively single-player game like *Spore* suggests that epic contexts combined with collaborative production tools and sophisticated content-sharing platforms can create opportunities for what we might call **lightweight, asyn-**

chronous **collaboration**. It's less immediately interactive, but it can still produce extreme-scale results. So far, *Spore* players from more than thirty countries have created and shared more than 144 million ecosystem objects, from creatures to buildings to space-faring vehicles.

OF COURSE, collaboration skills are on the rise around the world among non-gamers as well. From widespread basic Internet literacy and mobile technology smarts to rapidly expanding Web 2.0 and crowdsourcing know-how, people everywhere are becoming increasingly connected and improving their ability to cooperate, coordinate, and create together in many important ways. In this sense, gamers are just part of a larger social and technological trend toward more collaboration.

But gamers are having so much *fun* developing their collaboration skills, they're collectively spending more time than anyone else in the world honing and applying them. Every day and night, hundreds of millions of strangers from all over the world come together to prototype and playtest new ways of collaborating. The more they play together, and the closer they get to ten thousand hours of practice collaborating, the more justifiably optimistic they become about what they can accomplish together—and so they demand even more extreme collaboration challenges. And because gamers have developed such a growing appetite for collaboration at extreme scales, they've pushed the gaming industry to develop software and platforms that increasingly emphasize collaboration as a central gameplay mechanism.

As a result of the industry's relentless focus on innovating new ways to cooperate, coordinate, and cocreate, many online gamers are developing a new set of *collaboration superpowers* that transcend what they—and nongamers—are capable of doing in real-world, or nongame, environments. These gamers are on the front lines of testing and improving the ways we organize ourselves, amplify each other's individual abilities, and contribute to the common good.

Which gives us another fix for reality:

◯ FIX #13: TEN THOUSAND HOURS COLLABORATING

Compared with games, reality is disorganized and divided. Games help us make a more concerted effort—and over time, they give us collaboration superpowers.

What do I mean by collaboration *superpowers*?

A superpower is not just a new skill. It's a skill that so far surpasses any previously demonstrated skill, and it effectively changes our notion of what is humanly possible.

The term "superpower" suggests that something is happening outside the traditional model of learning and skill acquisition. Typically, we think of practice as moving us from a zero-skill level to basic competency and then, if we keep practicing, to proficiency and ultimately to mastery. But mastery presumes that there is some finite end to the skill level it is possible to achieve. So why stop at mastery? The term "superpower" reminds us that we are on the threshold of a new kind of capability, one that has not yet been mastered by anyone, anywhere. There's no telling yet how far these new capabilities will develop.

What, exactly, do these new capabilities look like?

In my research at the Institute for the Future, I've developed a model of how someone with collaboration superpowers works. It involves three key new skills and abilities.

Extraordinary collaborators are extremely extroverted or outgoing in a network environment—even if they're introverted or shy in face-to-face settings. They have what I call a **high ping quotient**, or high PQ. (In tech speak, a "ping" is a computer network tool that sends a message from one computer to another in order to check whether it is reachable and active. If it is, it will send back the message "pong," thus establishing an active line of communication.) Extraordinary collaborators have no qualms about *pinging*—or reaching out via electronic means—to others to ask for their participation. They're also highly likely to pong back when other people ping them. That's what makes a high ping quotient a form of social capital.

Of course, it helps to have a good sensibility about who to ping when. (Otherwise, you become a participation spammer.) That's why extraordinary collaborators develop a kind of internal **collaboration radar**, or sixth sense, about who would make the best collaborators on a particular task or mission. This sixth sense comes from building up a very strong social network and maintaining a kind of peripheral awareness of what other people are doing, where they are, and what they're getting good at. And it's not just an internal system: collaboration radar is often augmented with "ambient information systems," like Twitter lists, the Xbox 360 friends dashboard, or the Ground-crew volunteer availability system. The stronger your collaboration radar, the faster you can leverage individuals' abilities toward the right effort.

Finally, the most extraordinary collaborators in the world exercise a super-power I call **emergensight**. It's the ability to thrive in a chaotic collaborative environment. The bigger and more distributed a collaborative effort gets, the more likely it is to become both chaotic and hard to predict. We know this from physics and systems theory: bigger isn't more; it's different. That's the principle of *emergence*. It's impossible to predict what will happen at scale until you get there, and it's likely to be vastly more complex than you expected. Of course, with increased complexity comes increased potential for chaos.

Extraordinary collaborators are adept and comfortable working within complex, chaotic systems. They don't mind messiness or uncertainty. They immerse themselves in the flow of the work and keep a high-level perspective rather than getting lost in the weeds. They have the information stamina to filter large amounts of noise and remain focused on signals that are meaning-ful to their work. And they practice possibility scanning: always remaining open and alert to unplanned opportunities and surprising insights—especially at bigger scales. They are willing to bypass or throw out old goals if a more achievable or a more epic goal presents itself. And they are constantly zoom-ing out to construct a much bigger picture: finding ways to extend collabora-tive efforts to new communities, over longer time cycles and toward more epic goals.

These three ways of working make up what I consider to be the most important attributes of an extraordinary collaborator. Together, these traits enable us to discover and contribute our individual strengths and expertise to a large, open-ended effort.

These collaboration superpowers aren't widely distributed yet. They're concentrated among gamers who, for the past decade, have consistently played the games that have been on the leading edge of co-op, collective intelligence, and collaborative production. But, obviously, these collaboration superpowers would be extremely useful outside of game settings. They could be applied to tremendous effect across many different real-world domains: data collection and analysis, social action, risk assessment, scientific research, innovation of new products and services, and government, to name just a few.

Indeed, if these collaboration superpowers become sufficiently widespread, it's easy to imagine a future in which there is significantly more collective effort harnessed toward solving extreme-scale problems, like ending poverty, preventing catastrophic climate change, reducing terrorist activity, and improving global health. But before we can use these superpowers to solve real-world problems, we need to distribute them more broadly throughout society. Collaboration is most effective when there are diverse actors. We need to put these collaboration superpowers in the hands of as many people as possible — especially young people, who represent the next generation of social actors and problem solvers.

That's why, in my commercial game-design work, I am always drawn to projects that can serve as learning environments for collaboration superpowers. I think of these projects as global collaboration laboratories, or **collaboratories**: online spaces for young people from around the world to come together and test and develop their ability to cooperate, coordinate, and cocreate at epic scales.

The best way to understand the modus operandi of extraordinary collaborators is in the context of a real working collaboratory. So let's take a look at The Lost Ring, an alternate reality game that I designed for the 2008 Summer Olympic Games in Beijing.

THE LOST RING—A COLLABORATORY
FOR PRACTICING NEW SUPERPOWERS

The Olympic Games are broken. That's what I thought to myself in the summer of 2007 when I was first invited to direct an alternate reality game for the 2008 Summer Olympic Games in Beijing.

For 99.99 percent of the world, I thought to myself, the Olympics are all spectacle, a vicarious thrill at best. There's no real participation. No active engagement. We watch the games, but we don't actually get to play.

It wasn't supposed to be that way. The Olympic mission, after all, is to bring the world together through play. The Olympics are also meant to create global community. But even the biggest Olympic fans have virtually no interaction with other people from around the world during the games. We're not physically at the Olympic Village, where the many elite athletes congregate. Instead, we're at home watching the games on television. How can we expect to bring the world together through the Olympic Games if 99.99 percent of the world doesn't get to actually play?

This wouldn't have bothered me that much if I didn't actually believe in the Olympic mission. The modern Olympics are the best-known and longest-lasting effort to use games as a platform for establishing common ground, focusing global attention, fostering mutual regard, and creating global community. I couldn't imagine a better context than the Olympics for trying to build a global collaboratory.

That's when it occurred to me: could the Olympic tradition of bringing the world together for an intense period of play be extended from athletes to *gamers?*

If so, it would represent an ideal opportunity to give the growing generation of virtuoso gamers a chance to demonstrate their extraordinary talents to the world. Just like the world's greatest athletes, our best global gamers could show us collaborative feats previously unthinkable. They could inspire us all to push the limits of our own collaboration powers.

So, as early anticipation for the Olympics mounted a year in advance of the 2008 Beijing games, I accepted an invitation from McDonald's, a global

Olympic sponsor, and the International Olympic Committee to join the efforts of a leading digital creative agency in San Francisco, AKQA. Our shared objective was to create an online game that would give young adults around the world an opportunity to collaborate at a scale as awe-inspiring as the modern Olympic Games themselves. Together, we spent an entire year working with a creative development team of more than fifty people to help computer and video gamers turn the 2008 Summer Olympics into a game that *they* could play, a collaborative effort that *they* could undertake.

This is the story of The Lost Ring and how it reinvented the reality of the Olympic Games.

> *Most of you listening to this podcast will not believe the story I am about to tell you. How is it possible, you will ask, that the greatest sport of all time has been forgotten for almost 2,000 years? I'm Eli Hunt, and this is the legend of the Lost Sport of Olympia.*
> —from the Secrets of the Ancient Games podcast
> series, posted online February 24, 2008

> *The Ancient Greeks banned it, but we're playing it anyway!*
> —from an invitation to a Lost Sport of Olympia
> training event held in San Francisco, April 15, 2008

On February 24, 2008, a fictional character by the name of Eli Hunt launched a real podcast series called Secrets of the Ancient Games. The series was promoted by the International Olympic Committee on the home page of its highly trafficked website with the tag line "Investigate Olympic mysteries and learn about the history of the earliest games!" Visitors to Hunt's site discovered that the podcast series focused on the so-called Lost Sport of Olympia—a blindfold game that Hunt, an amateur archaeologist, believed the ancient Greeks had mysteriously banned from their Olympics before attempting to destroy all evidence that the game had ever existed.

Was the story of the Lost Sport of Olympia just an ancient urban legend? To prove otherwise, Hunt presented three compelling pieces of evidence: an ancient Greek pottery shard that depicted naked blindfolded runners; a defaced stone tablet dated to 530 BC and inscribed with training instructions for an Olympic sport requiring an odd combination of athletic skills—"trust, endurance, spatial memory, and orienteering"; and a twenty-one-hundred-year-old victory plaque for an ancient Olympic champion named Demetros—a name historians had never seen referenced on any other surviving Olympic artifacts. Even more mysteriously, the victory plaque called the unknown Olympian the "champion of *la paigna megas*"—the most important game.

Artist's illustration of the fictional "Pyron's shard," which
in the story of The Lost Ring dates to 740 BC and is said
to have depicted blindfolded Olympic athletes.
(AKQA, 2008)

If the blindfold game really was the most important ancient Olympic event, then why had historians never heard of it before? Hunt left his viewers with the following challenge:

> Was there really ever a lost sport? If so, how was it played? And why was it considered the most important of all ancient games? If the lost sport indeed existed, we can only assume that the ancient

Greeks themselves conspired to hide it from the rest of the world. But what would make them go to such lengths to conceal it? Even with my new research, it is difficult to accept that everything we think we know about the ancient Olympics may be wrong. But if I am right, and if the Greeks did hide the truth, then perhaps there are more clues out there for those of us who look closely enough . . .

Within twenty-four hours, a community of online gamers and bloggers caught wind of the lost-sport mystery and immediately took Hunt up on his challenge. They didn't necessarily believe his evidence — but they could sense that some kind of interesting game was afoot. And because Hunt subtitled his podcasts in seven different languages — English, French, German, Spanish, Portuguese, Japanese, and Mandarin — these gamers were from all over the world. Using a variety of translation tools to talk to each other, the gamers created a discussion forum, set up a chat room, assembled a wiki, and started e-mailing Hunt for more information. They were determined to get to the bottom of the mystery of the Lost Sport of Olympia.

The gamers followed a trail of clues from Hunt's online podcast to other blogs and websites, which created a vast web of real historical information and urban legend, and then finally to the real world, where over the span of just six short weeks they discovered twenty-seven physical artifacts: pages of a mysterious illustrated text called the Lost Ring Codex. Each page contained more information about the rules and the purpose of the lost sport. These pages, dated to 1920, were scattered across twenty-seven countries on five continents. The text was written in the universal language Esperanto and apparently had been created and hidden by an earlier generation of lost-sport investigators who had ultimately failed in their efforts to revive the ancient game.

How did they find these hidden artifacts? First they learned a real ancient Greek navigational shorthand known as "omphalos code," the subject of Eli Hunt's second podcast. Next, they worked together on a wiki to translate a set of twenty-seven omphalos codes into modern-day GPS coordinates. Then they called in favors from members of *their* social networks, who called in favors from members of their social networks, to find people willing to recover the pages of the codex in these far-flung real-world locations, from a bookstore in Johannesburg to a hostel in Rio de Janeiro to an art gallery in Bangalore.

As each page was recovered, players volunteered to translate them out of Esperanto into eight different languages—the game's original seven, plus Dutch, since a sizable contingent of players had formed in the Netherlands—and the gamers were thus able to create a new and complete record of the Lost Sport of Olympia for people all over the world.

The translated codex revealed, among other things, that the ancient game was officially called The Human Labyrinth, but nicknamed The Lost Ring. Piece by piece, players learned that it was a team sport for sixteen players: one blindfolded runner and fifteen "wall members." The wall members used their bodies to create a human-size labyrinth by standing on string laid out in the shape of an ancient Cretan maze. The runner, unable to see or feel his or her way, would try to escape from the center of the labyrinth as fast as possible, while the wall members hummed to help guide the runner. The official maze dimensions ensured that the teams never had enough members to cover the entire wall at any given time; instead, they would have to race ahead of the runner to create more wall before the runner reached that part of the labyrinth. Each labyrinth team competed against other labyrinth teams to get their runner out the fastest.

So what did the gamers do once they had solved the mystery of the ancient Olympic event? They did what any true gamers would do: they started to play it. And they committed to bringing back the lost sport in time for the real 2008 Summer Olympics. Teams formed across the world: in Singapore, Tokyo, Bangkok, and Shanghai; in London, Paris, Zurich, and Vienna; in San Francisco, Portland, New York, and Dallas; in Buenos Aires and São Paolo; in Johannesburg and in Cape Town; in Sydney, Melbourne, and Wellington.

Every weekend for several months, gamers gathered in different cities to revive the Lost Sport of Olympia and master the blindfolded labyrinth race. They uploaded hundreds of training videos to YouTube and added thousand of training photos to Flickr in order to show off how good they were getting, to teach other people how to play, and to trade strategies with other lost-sport athletes.

The teams got faster and faster as they developed better teamwork and tried out more and more complex strategies. Soon, each weekend a new world record was being set somewhere—Vienna this week, New York City another, Shanghai the next. The U.S. Olympic champion runner Edwin Moses volunteered to serve as the online virtual coach of the lost-sport athletes, sending weekly advice by e-mail and in online chats. Collectively, the lost-sport athletes worldwide egged each other on, improving their best times from an average of three minutes and thirty seconds per race when they started training to an average of fifty-nine seconds per race by summer's end, with a world's best of thirty-eight seconds.

A New Zealand lost-sport team posts a training video online.
(Still from video by Joshua Judkins, 2008)

On August 24, 2008, the closing day of the 2008 Summer Olympic Games, six months after Eli Hunt had uploaded his first podcast, one hundred of the world's best computer gamers put their lost-sport knowledge to the test. The six best human labyrinth teams in the world assembled in Beijing, San Francisco, Salvador da Bahia, London, Tokyo, and Wellington to compete in their own self-organized world championships. As the final medal counts were tallied for the real Olympic Games, these gamers-turned-athletes competed against each other for honorary gold (Tokyo), silver (San Francisco), and bronze (Wellington) medals in the Lost-Sport Olympics.

Those are just some of the highlights of The Lost Ring, an alternate reality game that took a full year to develop and another six months to play, eventually creating a player community made up of citizens from more than one hundred countries on six continents: 28 percent from North America, 25 percent from Europe, 18 percent from the Asia-Pacific region, 13 percent from Latin America, 9 percent from Oceania, and smaller clusters in areas such as Dubai, Israel, and South Africa. More than a quarter of a million gamers participated, and the most active participants—the core team of puzzle solvers, translators, social engineers, researchers, and athletes—numbered above ten thousand—just as many members as the community of official 2008 Olympic athletes. Together, those alternate reality Olympians created a new history of the games for an online audience of more than 2.9 million.[7]

FOR MCDONALD'S and AKQA, The Lost Ring was innovative marketing. Instead of delivering a typical television commercial about McDonald's support for the Olympics, they would create a branded game that gave players an opportunity to become actively immersed in the Olympics. For the International Olympic Committee, it was a chance to help make a long-standing tradition more relevant to the gamer generation.

For me, it was an opportunity to create a world-class global collaboratory. That's why every element of The Lost Ring was designed to challenge players to practice extraordinary cooperation, coordination, and cocreation.

To inspire global cooperation, we used the strategy of massively distributing game content in different languages, on localized Web communities, and across far-flung real-world geographic locations in order to make it impossible for any single country, let alone a single player, to experience the game alone.

Key online game clues were hidden on regional websites and social networks—for example, the social network Hi5, which is popular in Argentina; the video-sharing site 6rooms, which is popular in China; and Skyrock, a popular blogging community in France. And, of course, physical game objects were hidden in virtually every corner of the world. None of these clues or objects was redundant; each added an important piece of information to the history of the lost sport. Players therefore needed to work together to collect everything and translate it for other players. To do so, they had to radically expand their collaboration horizons, pinging their way through multiple levels of extended social networks to find people capable of showing up at the right place and the right time, or translating one language into another, so someone else could accomplish the next step in the chain that would lead them to their goal.

We also adopted the strategy of telling what gamers call a "chaotic story."[8] Instead of presenting players with a single means of consuming the game story, we broke it into thousands of pieces like a jigsaw puzzle and then diffused it across many different media platforms: podcasts and blog posts; videos and online photographs; e-mails and Twitter posts from game characters; even live instant message conversations and face-to-face interactions with characters portrayed by "game masters." This kind of chaotic storytelling mode forces players to actively make sense of the game content for themselves and for each other, using collective intelligence skills and collaborative authoring platforms. Until players put a chaotic story together, it doesn't really exist—it's just a web of evidence, the raw materials for a story. It's up to the players to do the actual final storytelling, which typically occurs on a wiki that ultimately represents an "official" story of the game.

In the case of The Lost Ring, the players pieced together the chaotic story and worked through the translation challenges primarily on a special Find the Lost Ring wiki they created. By the end of the game, it contained a total of 730 audio, video, and image files, as well as 943 articles—all created by players. The site's user-created pages included the Timeline, a detailed record of every major discovery and event during the six-month game; the Codex, a compilation of high-resolution scans of the twenty-seven manuscript pages found around the world, accompanied by their translation into nine additional languages; and Labyrinth Training Reports, where best times, videos, and other details of significant training events were recorded.

Of course, beyond the chaotic storytelling, the lost sport itself was designed to require intense collaboration. The rules of the game were cooperative—and they ensured that players would train locally as a team to improve their collective performance as a group, as opposed to competing locally with other athletes. For teams who became good at the game, the synchronized movement would provide a collective experience of flow similar to group dancing. And as in all of my live-action games, I snuck in some oxytocin-releasing touch—the wall members joining hands or gripping each other's shoulders—in order to strengthen the players' cooperative bonds. By making the human labyrinth together, they would be engaged in cocreating a peak experience.

But beyond the rules of the sport, what really required intense cooperative effort and coordination was the simple fact that no one had ever played this human labyrinth sport before. No one would know the rules and no one would be any good at it yet. Together, the participants would have to bootstrap their way to mastering the sport. They would have to work together as a global network to discover—to invent—the ins and outs of the sport for the very first time, to teach each other their best tricks, and to spread the secrets of the game online to as many people as possible. Players started scheduling their training sessions to maximize the number of time zones that could participate at the same time in normal waking hours. They spontaneously decided to stream video of their training sessions live via cell phone networks so that other cities' teams could watch and learn. These were extraordinary acts of coordination.

Finally, mass collaboration works only when everyone in the "mass" has something useful to contribute. Everyone needs to be given an opportunity to contribute from a position of personal strength—who they are and what they do best. So as I developed The Lost Ring as a collaboratory, one of my primary goals was to pioneer a system for helping players identify their own signature strengths, in order to help guide them to the kinds of contributions they could most effectively make in the game. That's why one of the centerpieces of the Lost Ring Codex was the legend of six ancient strengths, each named for an ancient Greek virtue, and each describing a distinctive way of contributing value to a group:

- *Sofia*: I bring wisdom, creativity, and cleverness to our mission. I am one of the knowledge seekers.
- *Thumos:* I bring courage, energy, and determination to our mission. I am one of the adventurers.
- *Chariton:* I bring heart, humanity, and charm to our mission. I am one of the connectors.
- *Dikaiosune:* I bring leadership, direction, and focus to our mission. I am one of the pilots.
- *Sophrosune:* I bring balance, self-control, and an open mind to our mission. I am one of the advisors.
- *Mythopoeia:* I bring optimism, vision, and artistry to our mission. I am one of the truth finders.

The codex included a twelve-question test to help you identify your primary and secondary signature strength. Players would choose from sets of statements the one that described them best. For example:

- I am an original thinker.
- I prefer to lead a life of adventure.
- I enjoy helping others.

And:

- I like being in charge.
- I am a fair and honest person.
- I always see the beauty around me.

(These choices respond to *sofia, thumos, chariton, dikaiosune, sophrosune,* and *mythopoeia,* respectively.)

Although the game presented these six strengths as a kind of ancient lore, in fact they are drawn directly from seminal positive-psychology research. In 2004, researchers Martin Seligman and Christopher Peterson published *Character Strengths and Virtues*, a manual with twenty-four such categories, divided into six groups: **wisdom and knowledge**—cognitive strengths that entail the acquisition and use of knowledge; **courage**—emotional strengths that involve the exercise of will to accomplish goals in the face of opposition, external or internal; **humanity**—interpersonal strengths that involve tending to and befriending others; **justice**—civic strengths that underlie healthy community life; **temperance**—strengths that protect against excess; and **transcendence**—strengths that forge connections to the larger universe and provide meaning.[9]

Together with the Values in Action (VIA) Institute on Character, Seligman and Peterson devised a 240-question inventory for measuring the positive emotional strengths that contribute to our success and well-being in life.[10] The goal of the inventory is "to help people evolve toward their highest potential," and it's the most scientifically validated test of personal character in the world. Yet many people have never heard of it, let alone taken it.

I wanted players to be able to find their signature strengths as members of the Lost Ring community because I believe that this positive-psychology resource can play an important role in creating ways for large numbers of people to contribute to a collaborative effort. So I modeled our ancient-strengths questionnaire on an abbreviated version of the official VIA inventory of strengths. My Olympic-themed twelve-question survey wasn't scientifically validated—and it didn't dive deep into the twenty-four strengths, just the six higher-order categories. But, as a first introduction to the strengths, I thought it would serve a powerful purpose: to help players start to identify their collaboration strengths and practice putting them to use in the Lost Ring mission.

Once they had completed the ancient-strengths test and had determined their primary and secondary strengths, players were invited to post strength badges on their social network pages, declaring, for example, "I am Sofia," with a description of what that meant. These badges became a visual cue for other players to start keeping track of others' strengths—in other words, they

were building up their collaboration radar. Eli Hunt and other game characters then began giving players game missions based on their strengths: for
example, the brainy *sofia* players were challenged to research little-known
facts about other games that really had been banned from the ancient Olympics, while the adventurous *thumos* players were given the task of going out
into the real world to hunt down the physical artifacts, and the highly social
chariton players were encouraged to be the social engineers of the game and
figure out how to extend the social network of the Lost Ring community.

Even the lost sport itself had special roles for every kind of strength to play:

- *Sofia*: You are the best engineers. Study the labyrinth plans—and
  arrive early to design and build the labyrinth.
- *Thumos*: You make the fastest runners. Get blindfolded and go
  for it!
- *Chariton*: You are the best coaches. Cheer on your team and trash-
  talk others.
- *Dikaiosune*: You make the best captains. Keep your team strong
  and focused on getting faster. Keep your wall coordinated and
  working together!
- *Sophrosune*: You make the best referees. Make sure everyone follows the rules. Keep time of the best scores.
- *Mythopoeia*: You tell the best stories. Take film and video of the
  game! And spread news of the best times from other cities—help
  your local team keep up-to-date on how the rest of the world is
  training.

For players trying to recruit more athletes to participate in the lost-sport
training events, and to engage the crowds of people who showed up to play
but hadn't been following the online story, the strengths test and assignments
proved an excellent resource. It helped the experienced players give prospec-
tive players a meaningful way to contribute right away. It gave them a tool for
directing new players to areas in which they were likely to experience success

and reap intrinsic reward—and it made sure that not one single potential contributor would find himself or herself without a satisfying task.

As Seligman and Peterson have often pointed out, we seem to be happiest when we are putting our signature strengths to good use in a group setting. The best evidence I've seen for this argument is when I watched Lost Ring players eagerly adopt their ancient-strength roles and perform them both online and in person as they brought the lost sport to life.

When we first launched The Lost Ring, we did not know where it would go. We gave the players the raw materials for staging their own collaborative effort—a series of online urban legends and mysterious physical documents that suggested the potential to revive a lost Olympic sport—but would they do it? And if so, how?

We were confident we could bring together a global community with our chaotic, multilingual narrative. But would the community actually bring the lost sport back to life? Would they invent their own ways to get not just good at the sport, but Olympic-athlete good at it? When planning the game, we had playtested the lost sport only a few times, mostly with the creative team for the project, and collectively we were very slow and very bad at it. We never imagined the athletic feats that our players eventually made themselves capable of—indeed, no one had ever imagined such a feat until the players undertook it. The lost sport had never really existed—and it never would have, either, if not for the concerted effort of the global gamers.

Edwin Moses, a real gold medal–winning Olympic athlete and multiple world-record setter in the 400 meter hurdles, sat down with our creative team to watch videos of the best players on each continent. He seemed genuinely impressed with the teams' performance and crafted individual video messages of support for each team. Later he answered players' questions in a streaming online broadcast, giving serious advice to them about how to best prepare for a gold-medal event. Eventually, he joined us at sunrise on the Great Wall of China, one of the official Olympic event sites, to coach in person the Beijing lost-sport team in our alternate reality gold-medal race. When we had first conceived of a blindfolded human labyrinth race, I never would have imagined that our players would take the game to such a level of athletic excellence, or that we would be able to genuinely engage and impress a real Olympic champion with the sport. But we did—thanks to the players' collaborative efforts.

In the end, our players produced two extraordinary collaborative results: a complete, extensive history of the ancient lost sport and its modern day revival on the Find the Lost Ring wiki—a 943-page multimedia document coauthored by more than a thousand of the game's leading players—and a community of athletes that made and raced labyrinths as if they had spent their entire lives (and not just six months) training for it. This was an act of true *emergensight* on the part of the players. From the complex, chaotic environment of Eli Hunt's legends and the scattered mysteries of the codex, they saw the opportunity to forge a clear, collaborative path together: to create an epic work of alternate reality history, and to stage this awe-inspiring six-continent spectacle. As a result, I count the lead players of The Lost Ring—particularly the thousand most active players who took ownership of the wiki and coordinated the months-long training in the lost sport—among the true collaboration virtuosos of their generation.

We are all born with the potential to develop collaboration superpowers. Scientific research shows that we have both the ability and the desire from early childhood to cooperate, to coordinate activity, and to strengthen group bonds—in other words, to make a good game together. But this potential can be lost if we don't expend enough effort practicing collaboration.

Fortunately, we have many collaboratories for doing so already. In addition to global alternate reality games like The Lost Ring, any good online game with co-op mode, collaborative production opportunities, and a thriving Wikia culture, for example, provides the perfect opportunity to practice collaboration superpowers. And thanks to the increasing availability of good games worldwide, we will have more and more opportunities than ever before to develop these superpowers.

This is increasingly true even in developing countries, which traditionally have had limited access to leading-edge online games and game platforms. Today, game developers are creating online game platforms specifically for the technology constraints of emerging technology markets like India, Brazil, and China. For example, the lower-priced game console Zeebo, which describes itself as "the video game console for the next billion," connects low-energy-demand gamer consoles via mobile phone networks rather than broadband Internet. Meanwhile, networked games are being developed for the mobile phones that are ubiquitous even in the most isolated villages across Africa.

As the game industry continues to emphasize co-op, collective intelligence, and collaborative production modes of play, collaboration superpowers will spread more widely throughout gamer culture. And as more and more people start to think of themselves as gamers—perhaps in no small part because they want to develop their own collaboration superpowers—these extraordinary new skills and abilities will become ordinary—the norm rather than the exception.

So what can we do with the collaboration superpowers we develop over the next decade and beyond? One of the first epic goals for gamers worldwide may be simply to survive the twenty-first century.

In their 2006 book *Wikinomics*, the breakthrough manual for extreme-scale collaboration in the real world, Don Tapscott and Anthony Williams famously implored: "We must collaborate or perish—across borders, cultures, disciplines, and firms, and increasingly with masses of people at one time."[11]

"Collaborate or perish" is perhaps the single most urgent rallying cry for our times. The ability to collaborate at extreme scales isn't just a competitive advantage in business or in life anymore. Increasingly, it's a survival imperative for the human race. As the *Wikinomics* authors suggested several years later in an updated preface to the book, "The killer application for mass collaboration may be saving the planet, literally."[12]

A killer application is a program so valuable, it proves the core value of the larger system and drives massive amounts of people to adopt it; e-mail, for instance, was considered the killer app for home Internet access. I believe wholeheartedly that the core value of developing our collaboration superpowers will be proven by games that help gamers save the real world—by changing how we consume energy, how we feed ourselves, how we create better health, how we govern ourselves, how we conceive of new businesses, and how we take care of each other and the environment.

But these fundamental changes don't happen overnight—surviving the twenty-first century together will require us to adopt longer horizons of thinking, acting, and collaborating. We need to play games that stretch our collective commitment months, years, or even decades ahead.

We need to start playing with the future.

CHAPTER FOURTEEN

Saving the Real World Together

We are living in a geological era that scientists dub the "anthropocene epoch," from the Greek *anthropo-*, for "human," and *-cene*, for "new" or "recent." It's the age of human impact on the earth.

Our impact is measurable in myriad ways: increasing levels of carbon dioxide in the atmosphere, deforestation and continental erosion, a rising sea level. We may not have set out to remake the planet in any of these ways—but we have nonetheless. And now we must learn better ways of remaking it, this time with intention, discipline, and purpose.

As Steward Brand, author of *Whole Earth Discipline*, puts it, "Humanity is now stuck with a planet stewardship role. . . . We are as gods and *have* to get good at it."[1]

Brand is perhaps best known as the founder of the *Whole Earth Catalog*, a countercultural catalog of "tools and ideas to shape the environment" published from 1968 to 1972. (When he launched that catalog, he wrote, "We are as gods and might as well get good at it.")[2] In 1996 he cofounded the Long Now Foundation, a San Francisco–based foundation dedicated to long-term thinking and responsibility—for the earth, and for the survival of the human species—over the next ten thousand years and beyond. If we want to stay on this planet for anywhere near that long, Brand says, we have to become better at strategically affecting our ecosystem. "We are forced to learn planet craft— in both senses of the word. Craft as skill and craft as cunning." We not only have to master the ability to change how our ecosystem works, we also have to figure out the right ways to change it. And that won't be easy.

"The forces in play in the Earth system are astronomically massive and un-imaginably complex," Brand writes.[3] "We're facing multidecade, multigenera-tion problems and solutions. Accomplishing what is needed will take diligence and patience—a sustained *bearing down*, over human lifetimes, to bridge the long lag times and lead times in climate, biological, and social dynamics."[4]

Fortunately for all of us, gamers actually have a head start on this mission.

Gamers have been mastering the art of planet craft for years. There's actually a genre of computer games known as "god games"—world- and population-management simulations that give a single player the ability to shape the course of events on earth in dramatic ways, over lifetimes or longer.

As we've seen, Will Wright's *The Sims* gives players godlike powers over the daily lives of individual people. Sid Meier's *Civilization* challenges players to guide a civilization (such as the Aztecs, the Romans, the Americans, the Zulus) from the start of the Bronze Age, six thousand years ago, through the Space Age, or AD 2100. And Peter Molyneux's *Black & White* invites players to govern the entire biome of a remote island, inspiring either joyful worship or terrified obedience in the island's tribal population by performing a com-bination of benevolent and evil divine ecological interventions.

What all of these god games have in common is that they encourage play-ers to practice the three skills that are critical for real planet craft: taking a long view, ecosystems thinking, and pilot experimentation.

Taking a long view means working at scales far larger than we would ordi-narily encounter in our day-to-day lives. Players of god games have to consider their moment-by-moment actions in the context of a very long future: an entire simulated human life, a single civilization's rise and fall, or even the entire course of human history.

Ecosystems thinking is a way of looking at the world as a complex web of interconnected, interdependent parts. A good ecosystems thinker will study and learn how to anticipate the ways in which changes to one part of an eco-system will impact other parts—often in surprising and far-reaching ways.

Pilot experimentation is the process of designing and running many small tests of different strategies and solutions in order to discover the best course of action to take. When you've successfully tested a strategy, you can scale up your efforts to make a bigger impact. Since god gamers want to maximize their success, they don't just come up with one plan and stick to it. Instead, they carefully feel their way around the system, poking and prodding until they find the strategies that seem to reliably maximize success.

Taken together, these three ways of thinking and acting are exactly the kinds of effort Brand recommends in *Whole Earth Discipline*. Instead of seizing the day, he says, "Seize the century."[5]

He advises, "Participation has to be subtle and tentative, and then cumulating in the right direction. If we make the right moves at the right time, all may yet be well."[6]

OF COURSE, we can't actually use existing commercial computer games as test environments to solve the real problems we face. They radically simplify the forces at play in the complex ecosystems we live in. But as we try to develop systems for engaging massively many people in world-changing efforts, we can take an important cue from the most successful god games. Specifically, we can learn from their ability to change the way players think about the world, and their own powers within it.

Take, for example, the most epic god game yet designed—the universe simulation *Spore*, developed by Will Wright and produced by Maxis Software. Of all the god games to date, *Spore* is the most explicitly linked to the notion of planet craft—and the most intentionally focused on getting players to think of themselves as capable of changing the real world.

In *Spore*, players control the development of a unique species through five stages of evolution: from single-cell origins (stage one) into social, land-dwelling creatures (stage two), who form tribes (stage three), build techno-logically sophisticated civilizations (stage four), and ultimately venture off into intergalactic space exploration (stage five). Each stage zooms out to give the player control over a more complex system. Players advance from manipulating cellular DNA to increasing their creature's intelligent behaviors; from organizing a division of labor in their tribe to growing a global economy; from advancing national interests through trade, military action, or spiritual outreach to colonizing other planets and transforming them into inhabitable ecosystems. They can spend as much time as they want in any stage, piloting different strategies for improving their species and transforming the environment.

The game is fun and rewarding to play, but it's meant to accomplish more than just relieving boredom or making us happy. As Wright has said on numerous occasions, the game is meant to spark a sense of *creative capability* among players, and to inspire them to adopt the kind of long-term, planetary outlook that can save the real world.

Consider this exchange, which occurred shortly after the 2008 release of *Spore*, when the popular science magazine *Seed* hosted a public salon between Wright and Jill Tarter, a noted astrobiologist. The topic of the salon: how games like *Spore* are preparing young people to take a more active role in reimagining the real world.

> **TARTER:** I keep thinking about the generation that's getting exposed to all this wonderful, rich opportunity of game playing as education, and that they expect to be able to manipulate the real world the way they do the game world. How do we bridge that? How do we turn them into socially functioning members of humanity on one planet? [. . .]
>
> I'm eager to understand how learning to be good at a game makes you good at life, makes you good at changing the world, and gives you skills that are going to allow you to reinvent your environment.
>
> **WRIGHT:** Well . . . if there's one aspect of humanity that I want to augment, it's the imagination, which is probably our most powerful cognitive tool. I think of games as being an amplifier for

the imagination of the players, in the same way that a car amplifies our legs or a house amplifies our skin. [. . .]

The human imagination is this amazing thing. We're able
to build models of the world around us, test out hypothetical
scenarios, and, in some sense, simulate the world. I think this
ability is probably one of the most important characteristics of
humanity.[7]

Why does Wright believe that augmenting our natural capacity for imagination is so important at this precise moment in human history? It's a matter
of survival, pure and simple.

The name *Spore* is itself an important clue: the definition of a spore, in
biology, is "a reproductive structure that is adapted for dispersal and surviving
for extended periods of time in unfavorable conditions."[8] It's a perfect metaphor for the present circumstances of the human race.

We have collectively entered into what is all but certain to be a time of
increasingly unfavorable planetary conditions, largely of our own making—an
unstable climate, extreme weather, and an increasingly depleted environment. We need to adapt for survival. We need to imagine planetary-scale solutions and disperse them as far and wide as possible.

We need to become like spores ourselves.

And there's an explicit call to action to do so, for players who complete all
five stages of the game successfully. *Spore* has what game developers call a
"primary win condition": a supergoal that represents the ultimate achievement in the game. The primary win condition in *Spore* is to develop your
single-cell creature into such a successful intergalactic space-faring civilization that it eventually reaches one galactic destination in particular: a super-
massive black hole at the center of the galaxy.

Players who reach the black hole receive a "staff of life," which allows them
to transform any planet in the *Spore* galaxy into a vibrant, diverse ecosystem:
teeming with plants and creatures of all kinds, with breathable atmosphere,
sustainable food webs, and plentiful water supply. (No wonder players also
refer to it as the "Genesis device.")

The staff of life is a shortcut to making an otherwise uninhabitable planet inhabitable. Along with the staff of life, players receive a special message and mission:

> You have traveled very far and overcome many obstacles. Your creative efforts have not gone unnoticed. Your heroic efforts have proven you deserving, worthy of advancement to the next level of your existence. You are now to be given the power. Yes, that's right, THE POWER. The power to create and spread life, intelligence and understanding throughout the cosmos. Use this power wisely. There is a wonderful opportunity to start on one particular planet: Look for the third rock from *Sol*.

Sol is Latin for "sun," and so the *Spore* community has translated this final message from the game as a playful imperative to remake our own Earth — which is, of course, the third rock from our sun.

In the end, a win in *Spore* is a push back to the real world. Players are told, "Your gameplay has prepared you to become a real creator and protector of life on Earth." Not by making them an expert in geoengineering, atmospheric science, or ecological planning, certainly, but rather by creating the seed of planetary creativity and activism. As Wright said at the *Seed* salon:

> All of the really tough problems we're facing now are planetary problems. There's real value in being pushed toward global awareness and looking long-term. That's one of the things that I find very useful about games. . . . I think these are the timelines we need to be looking at — the one-hundred- or two-hundred-year horizons. Because most of the really bad stuff that's happening right now is the result of very short-term thinking.

We can break free of the cognitive chains of short-term isolated thinking, with games that direct our collective attention to the future and challenge us to take a global perspective.

––––––

GOD GAMES LIKE *Spore* have gotten us successfully started on this journey. Now a different genre of games can get us where we need to go: massively multiplayer forecasting games, or **forecasting games** for short.

Forecasting games combine collective intelligence with planetary-scale simulation. They ask players to reimagine and reinvent the way we feed ourselves, the way we transport ourselves, the way we get water, the way we design cities, the way we manufacture everything, the way we power our lives. They're designed to create diverse communities capable of investigating the long-term challenges we face, propose imaginative solutions, and coordinate our efforts to start putting our best ideas into action at the planetary scale.

It's a process I call **massively multiplayer foresight**. And future-forecasting games are the perfect tool for helping as many people participate in the process as possible.[9]

Which brings us to our final fix for reality:

↻ FIX #14: MASSIVELY MULTIPLAYER FORESIGHT

Reality is stuck in the present. Games help us imagine and invent the future together.

How exactly does massively multiplayer foresight work? The best way to understand the process is to start with the project that inspired the forecasting game genre.

World Without Oil: Play It Before You Live It

You know it's bad for you.
You'll cut back someday.
On April 30, join a World Without Oil—and play it before you live it.
　　　　　　　　　　　　　　　　—Announcement for the game

In April 2007, the world ran out of oil.

It didn't run completely out of oil—it simply ran out of *enough* oil. The daily demand for oil worldwide began to outpace our daily production capacity. Shortages broke out, reserves were tapped, and yet the gap between global supply and demand grew wider.

The United States was among the hardest countries hit. During the darkest days of the crisis, as many as 22 percent of Americans were unable to get access to gas, while one in ten U.S. companies buckled under the strain of higher fuel costs and diminished operating capacity.

Two main strategies emerged to deal with the oil crisis.

We could collectively reduce our daily demand for petroleum in order to create equilibrium with the available supply.

Or we could compete even more aggressively for the available oil—with our own individual neighbors, with other companies, with other states, and with other countries.

Of course, this didn't really happen—at least not for most of us.

But for two thousand online gamers, this peak-oil scenario was the basis for a life-changing six-week experiment: a collaborative simulation designed to find out what *would* happen if demand for oil did eventually outstrip our supply, and what we could collectively do about it.

The project was called World Without Oil (WWO), and it was the first massively scaled effort to engage ordinary individuals in creating an immersive forecast of the future.

 HOW TO PLAY WORLD WITHOUT OIL

At heart World Without Oil is very simple. It's a "What if?" game.

What if an oil crisis started today—what would happen? How would the lives of ordinary people change?

What would you do to survive the crisis? How would you help others?

Let's play "What if?" and find out.

Create your own story of life during the oil crisis—and share it with us by e-mail or by phone call, by photos or by blog post, by videos or podcasts.

Then join our citizen "nerve center" at worldwithoutoil.org to track events and share solutions. Every day, we'll update you with news about the crisis, and highlight our favorite stories from across the country and around the world.

No expert knows better than you do how an oil shock could impact your family, your job, your town, your life. So tell us what you know.

Because the best way to change the future is to play with it first.

Funded by the Corporation for Public Broadcasting and presented by the Independent Television Service (ITVS), World Without Oil was first conceived by Ken Eklund, an independent writer and interactive developer based in San Jose, California. He pitched the idea in response to an ITVS announcement of funds up to $100,000 being made available for innovative educational online games. I was invited by ITVS to serve on the evaluating committee for online game proposals.

"No one today has a clear picture of oil availability or what will happen when demand inevitably outstrips supply," Eklund wrote in his proposal. "That will largely depend on how well ordinary people respond to the crisis. Until now, no one has ever thought to ask them what they might do. WWO will evoke the wisdom of crowds in advance, as players work together to gain grassroots insights into the forces that will rule at street level in a crisis—and figure out the best ways to prepare, cooperate, and collectively create solutions if and when a real peak-oil shortage happens."

It was designed as a massively multiplayer thought experiment: players would spend six weeks imagining how such a crisis might play out in their local communities, their industries, and their own lives. They would make highly personal forecasts using online social media. And they would rely on an "alternate reality dashboard" to get daily updates on the scenario, in the form of fictional news stories, video reports, and economic indicators from the peak-oil crisis in order to flesh out their personal forecasts in more detail.

Players would also be strongly encouraged to take the simulation a step further, and spend some time each day living their real lives as if the simulated oil shortage were true. How hard would it be to get to work, or to prepare dinner, or to see friends and family if the fictional simulation were real? Players were challenged to test their own ability to adapt, rapidly and dramatically, to a potential oil crisis. Instead of just imagining a peak-oil scenario, they could start making changes and testing adaptive solutions for real.

Each day in real time would represent a week in the simulation. This would enable players to consider longer-term impacts and strategies. The game itself would last for thirty-two days, so the scenario could play out over thirty-two weeks.

WWO would give players firsthand insight into a plausible future, helping them prepare for, or even prevent, its worst outcomes. The game would also create a collective record of how a real peak-oil scenario might play out—a kind of survival guide for the future, a record of tremendous value for educators, policy makers, and organizations of all kinds.

I happily accepted Eklund's invitation to join the project team as the game's "participation architect"—a fancy way of saying my job was to help make sure every single player found a way to contribute meaningfully to the collaborative effort.

Of course, to start, we had to attract a community of players. I set our target at one thousand players, a number based on my experience with online communities and collective intelligence. One thousand participants seems to me to be a critical threshold to allow for an online game to get interesting—to ensure enough diversity among players, to have enough participants to tackle missions on an epic scale, and to produce enough chaotic interaction to generate complex and surprising results.

For six weeks before we launched, we spread the word online and at public events. We asked our friends and colleagues to blog about it. I announced the game in my keynote for the Serious Games Summit, an annual two-day meeting in San Francisco for people working on games designed to teach, train, and solve real problems. ITVS reached out to educators and media creators across the country. There wasn't any other marketing plan or promotional budget for the game. It was simply an open, public invitation to simulate the future, and the game was free to play.

So who showed up to play? They numbered just over nineteen hundred (nearly doubling our initial goal), evenly divided between men and women, and representing all fifty United States and a dozen countries abroad. Most players were in their twenties or thirties, but there were notable clusters of every age group, from teenagers to seniors. And our most active players brought together an astonishingly diverse range of personal concerns and real-life expertise to the game. For example:

- Peakprophet, a self-described "hobby farmer" in Tennessee, who forecast the collapse of the fresh-food supply chain—and then took it upon himself to train other players how to grow their own food and increase their food self-sufficiency.
- Lead_tag, a soldier stationed in Iraq, who blogged every single day of the game, creating a series of thirty-two reflections on the challenges of fighting a war during an oil crisis.
- Anda, a college student pursuing a bachelor of fine arts in graphic design at the San Francisco Art Institute, who created a series of eleven Japanese manga-style Web comics about how she and her friends would help each other during the oil crisis, and how it might affect their ability to find work after graduation.
- OrganizedChaos, a dispatcher at a General Motors plant in Detroit, who contributed fifty-five blog posts, videos, and podcasts, and found herself forecasting that pretty soon—peak oil or not—she would no longer have a job. As a result, at the end of the game she decided to go back to school in real life to prepare for a new career in a postoil economy.

Once we'd assembled our forecasting community, it was crucial for us that a significant portion of our players stay engaged with the game for its entire six-week duration. That's because when it comes to future forecasting, our first ideas are often the most obvious and generalized, and therefore the least useful. It takes a while, even for an experienced forecaster, to drill down to the most interesting specifics and spin off unexpected possibilities. So we adopted several strategies to keep players engaged and actively investigating different aspects of the scenario.

First, each game day we added a new piece of information to the mix: rolling brownouts from oil-dependent power companies; airlines canceling flights and dramatically raising the cost of tickets; empty shelves and food shortages due to inability of deliveries to be made to local stores. In return, players told us about difficulties dealing with unreliable power at home; business travelers getting stranded in other countries when airports unexpectedly shut down; public transportation overcrowding in towns and cities with previously underutilized systems; a disruptive uptick in work-from-home days; the rise of bicycle thefts and a new bicycle black market; impromptu homeschooling as a result of gas shortages in suburban and rural areas; and neighborhood potluck meals to deal with the food shortage.

Another important tool for continuing participation was our alternate reality dashboard, which included a map depicting thirty-eight different regions, such as the Boston metro area, the Cincinnati–Columbus metro area, the Great Lakes, the High Plains, and the Atlantic South, each with its own set of "power meters" reflecting the local rise and fall in quality of life, economic strength, and social stability. The power meters fluctuated in direct response to player activity. The more positive forecasts they made, the more cooperative strategies they developed, and the more actively they reduced their own collective daily oil consumption, the more favorable the regional metrics. However, if players chose to imagine a darker turn of events, or if they chose to focus on how increased competition might play out, or if they reported significant difficulties or hardships in adapting to a lower-consumption lifestyle, the metrics would reflect increased chaos, rising misery, or even economic collapse. The meters created a clear feedback loop between players' stories and the scenario updates.

Of course, the sizable online audience that assembled for World Without Oil was also a huge incentive for players to tell the best stories possible. For every active forecaster, we had an additional twenty-five people watching the game and writing about it. This amplification of their ideas helped make the players' efforts feel more meaningful.

In the end, the game produced more than a hundred thousand online media artifacts—including a core set of more than two thousand future-forecasting documents from the players and tens of thousands more blog posts and articles reflecting on the game and its findings. One reviewer called it a "huge growing, twisting network of news, strategy, activism, and personal expression."[10]

At first, the majority of players focused their efforts on imagining how local, regional, and international competition for oil resources would play out in this new environment of increased scarcity. They exercised a dark imagination, anticipating the worst possible outcomes and the most serious threats. They documented gas theft, riots, food shortages, widespread looting, job loss, school closures, and even military actions worldwide. At a more personal level, they told stories of personal stress, anxiety, and families in crisis.

But over the course of thirty-two weeks, the balance shifted. About halfway through the game, having exhausted their dark imagination, players began focusing on potential solutions. They started imagining best-case-scenario outcomes: new ways of cooperating to consume less, a focus on local com-munity and neighborhood infrastructure, less time spent commuting, the geo-graphic reassembly of extended family, and more time spent in pursuit of a new American dream—happiness built around notions of sustainability, sim-plicity, and stronger social connectivity.

The game started with near-apocalyptic undertones; it ended with explicit, if cautious, optimism. The best-case-scenario outcomes were posed not as probabilities—and certainly not as inevitabilities—but rather as *plausible* pos-sibilities worth working toward.

There was no explicit prompt to start with dark imagination and only later veer toward optimism. But it is, in fact, a very sound forecasting strategy.

Researchers have pointed to a particularly American failure to believe that the worst can really happen, because we're systematically trained by our culture to focus on the positive. It's a failure that makes us more susceptible to catastrophic events, like Hurricane Katrina or the 2008 housing market collapse, for example. In *Never Saw It Coming*, sociologist Karen Cerulo argues that our collective inability to focus on negative futures is our culture's biggest blind spot.[11] As one reviewer of Cerulo's book summed it up: "We are individually, institutionally, and societally hell-bent on wishful thinking."[12] We are very good at positive thinking, but we tend to avoid articulating worst-case scenarios, which unfortunately makes us more vulnerable to them and less resilient if they occur.

World Without Oil gave players a space for nonwishful thinking; that's what created a sense of urgency to find solutions. That mind-set also lent a sense of gravitas and realism to even the most hopeful stories players' told later in the game—stories we later compiled into a guide, "A to Z: A World Beyond Oil."[13] It contains some of the most interesting community solutions players devised and can give you a taste of how massively multifaceted the final collaborative forecast was. Here are a few of my favorite topics from the document:

- Architecture Without Oil—notes from attendees of a national architecture convention on how to design and build homes for a world without oil
- Fellowship Without Oil—a collection of sermons and prayers from pastors, ministers, and other spiritual leaders offering guidance for how to act compassionately during the oil crisis
- Neighborhood Without Oil—guidelines for how to build stronger personal relationships with our geographically closest neighbors, the people most likely to be of assistance to us during an oil crisis
- Your Mama Without Oil—reflections from mothers of young children on how to parent in a world without oil
- Zoom Zoom Without Oil—conversations among automotive rac-

ing fans about the future of NASCAR and potential partnerships with alternative vehicle races, including electric vehicle races and human-powered vehicle races

"The forecasts are of astonishing quality," one reviewer said of WWO afterward. "The players got to the heart of a complex subject."[14] I think "heart" is a key word here, because players were telling stories about the futures they cared about most—the future of their industry, their religion, or their own town and their children.

After the game, creative director Ken Eklund reflected on what it had accomplished. "WWO didn't only raise awareness about oil dependence. It roused our democratic imagination. It made the issues *real*, and this in turn led to real engagement and real change in people's lives. Via the game, players made themselves better citizens." This is clearly evident in what players reported afterward. One player reported:

> I really mean it when I say WWO changed my life. I really have been using my cloth bags at the stores, walking more/driving less, turning off lights, and, yes, recycling. My friends, family and co-workers have all noticed the difference. In all seriousness, this entire thing has made me a different person.[15]

While another wrote:

> This experience has been just incredible for me. I've learned so much and started to think about even small things in my daily life in new ways. . . . Your stories and suggestions give me hope, that good ideas are emerging, that people are reaching out to help each other through these times, that necessary skills and knowledge are being saved and treasured for times when we will need them desperately. You show me that many really great people are out there . . . [and] you'll lead the way through.[16]

Today, the entire simulation has been preserved in a sort of online time machine at Worldwithoutoil.org, where you can experience the game from day one all the way through day thirty-two. Each day of gameplay is captured in time so you can see exactly how the collaboration unfolded; there are also guidelines for playing the game yourself today—on your own, with your family, with colleagues, with a classroom, or with your neighbors. In fact, the simulation has been repeated many times at a smaller scale to help individuals and communities prepare themselves, and invent their own solutions, for living in a world beyond oil.

PERHAPS YOU'RE WONDERING — as many people have asked me since—why did the players participate?

Why would anyone want to play a serious game like World Without Oil instead of a fantasy game, an escapist game, a completely feel-good game?

I asked myself the same question—before WWO launched, while it was being played and afterward, even when we had proof that players were enjoying themselves and audiences found it compelling to watch the project unfold.

Here's what I've come to believe about a game like WWO. By turning a real problem into a voluntary obstacle, we activated more genuine interest, curiosity, motivation, effort, and optimism than we would have otherwise. We can change our real-life behaviors in the context of a fictional game precisely because there isn't any negative pressure surrounding the decision to change. We are motivated purely by positive stress and by our own desire to engage with a game in more satisfying, successful, social, and meaningful ways.

I also firmly believe that many gamers want to do something that matters in the real world as much as their efforts matter in the game world. One player summed this up best:

> Looking back at World Without Oil, I think it is the most amazing, best multiplayer game I have experienced. Usually gaming takes time away from accomplishing useful things in real life, but WWO

taught me a lot, lowered my electric bill, and got me focused on doing things that matter to me.[17]

Gamers are ready and willing to take on challenges outside of strictly virtual environments. Meanwhile, people who don't ordinarily play games are happy to do so when it can help make a difference in the real world.

The numbers are still small. Two thousand players doesn't begin to compare with *Spore*'s active community of more than a million. But unlike *Spore*, which represents roughly two decades of some of the smartest computer programmers, some of the most creative game designers, and some of the most brilliant artists in the world working together to advance the genre of planetary simulation and god games (*SimEarth* was released in 1990), we are essentially still in the *Pong* days of future-forecasting games. (With the operating budget of *Pong* to boot.)

We are *Pong*, competing with *Spore*. It's not much of a matchup yet.

But as the field attracts more of the world's best programmers, storytellers, designers, and artists, as more people are exposed to these games and learn how to play them, and as we invest millions, rather than thousands, of dollars in developing these future worlds, we will grow our future world–building skills just as we've grown our virtual world–building skills over the past thirty years. With enough attention and investment, we will start to create immersive future environments as engaging as our favorite virtual worlds.

WORLD WITHOUT OIL changed the lives of many of our players—but it was also a life-changing experience for me.

It was the proof-of-concept game that convinced me we really can save the real world with the right kind of game. It's the project that inspired me to define my biggest hope for the future: that a game developer would soon be worthy of a Nobel Prize.

I've since taken to advertising that goal everywhere I go, in the hopes of inspiring other game developers to join me in my mission. Of course, both inside and outside the game industry, when I suggest the idea, I'm often met with skepticism. How could a game possibly accomplish enough real-world good to warrant such a prize?

Even on the heels of a project as promising as World Without Oil, it's true that winning a Nobel Prize is a fairly bold ambition. But consider this: Albert Einstein, who won his own Nobel Prize in physics in 1921, once famously said, "Games are the most elevated form of investigation." This quotation appears in multiple biographies of Einstein and circulates widely in various collections of famous sayings, but, interestingly, its origins remain elusive. No one seems to have recorded the context of Einstein's statement—when or where he said it, or what he meant by it. Why would an esteemed physicist call *games*, and not *science*, the most elevated form of investigation? It's an unsolved mystery—one I've spent much of my free time puzzling over.

Although of course I can't say for sure I've solved the mystery, I do have a theory. And it comes directly from working on World Without Oil.

Einstein, we know from many biographers, was a gamer—albeit a sometimes reluctant one. He had a lifelong love-hate relationship with the game of chess. He played it enthusiastically as a child, although he gave it up for much of his adult life, even once insisting to the *New York Times*, "I do not play any games. There is no time for it. When I get through work I don't want anything which requires the working of the mind."[18] Yet many friends and colleagues recall playing countless games of chess with Einstein, particularly later in his life.

Historians have suggested that Einstein avoided chess during the height of his scientific career precisely because he loved it so much and found it so distracting. "Chess holds its master in its own bonds," he once said, "shackling the mind and brain."[19] In other words, when he started thinking about chess, he found himself unable to stop. Why? Most likely because, as so many chess masters have noted, the game is an incredibly compelling problem that becomes more compelling the longer you think about it.

The central problem of chess is perfectly constructed, clear, and constrained: how do you manipulate a set of sixteen resources of different abilities in order to capture your opponent's most valuable asset, while simultaneously protecting your own? But it can be approached with endlessly many different strategies, each strategic effort changing the future possibilities in the problem space. As one famous chess saying goes, "Chess is infinite."

> There are 400 different positions after each player makes one move apiece. There are 72,084 positions after two moves apiece. There are 9-plus million positions after three moves apiece. There are 288-plus billion different possible positions after four moves apiece. There are more potential games than the number of electrons in our universe.[20]

The possibility space of chess is so massive and complex, one individual has no hope of understanding or exploring it fully—even if one spends a lifetime, as many chess players do, investigating it.

Fortunately, while chess is a two-player game, it is also a massively multi-player project. The global community of chess players has collaborated for centuries to explore and document its problem space as thoroughly and imaginatively as possible. Indeed, for as long as modern players have played chess, they have recorded their games, shared strategies, formalized successful approaches, and published them for others' benefit. Even after centuries of collective play, the chess community continues to seek a better understanding of the problem, to invent more surprising and successful approaches, and to hold the massively many possibilities of the game in their head as they drive the sum human understanding of the game forward, one move at a time.

To play chess as a more than casual player is to become a part of this problem-solving network. It means joining a massively collaborative effort to become intimately familiar with an otherwise unfathomably complex possibility space. And that's what I believe Einstein meant when he described games as an elevated form of investigation. When enough people play a game, it becomes a massively collaborative study of a problem, an extreme-scale test of potential action in a specific possibility space.

I believe that's the direction we're heading with forecasting games. These games help us identify a real-world problem and study it from massively multiple points of view. They present the problem in a compelling way, and they help us compile a record of massively multiple strategies for addressing it. They give us a safe space to play out the possible consequences of each and every possible move we could make. And they help us anticipate the massively multiple moves that others could conceivably take.

That's actually what we tried to do with World Without Oil. We defined a problem: an oil shortage, with no available means of increasing supply. There was a clear goal: to resolve the imbalance between supply and demand. The possible strategies were infinite. We asked players to craft their own strategies, based on their own unique points of view: a combination of location, age, life experience, and personal values. We asked them to test, on a local scale, different actions, and to report their findings. Taken together, all the players' stories and solutions represent massively multiple perspectives on the same problem. It was a truly elevated investigation.

And in a world of changing climate, geopolitical tensions, and economic instabilities, there are plenty more problems to be tackled with our collective imagination.

If we can develop the same kind of intelligence about the real problems we face as players do about their favorite games, then we *will* be able to practice better planet craft. We'll elevate our collective understanding of the challenges we face. And we'll build a global community of individuals ready to play a role in discerning the right moves to make in the future.

At the end of World Without Oil, I was struck by how optimistic players were. Despite having spent nearly a month imagining incredibly dark forecasts, our players wound up feeling better—not worse—about the future and their ability to impact it. They experienced a sense of improved capability, greater resilience, and realistic hope.

In other words, they became what futurist Jamais Cascio calls "super-empowered hopeful individuals," or SEHIs.[21]

A **SEHI** (pronounced *SEH-hee*) is someone who feels not just optimistic about the future, but also *personally capable* of changing the world for the better. According to Cascio, SEHIs get their confidence from network technologies that amplify and aggregate individual ability to impact the common good.

Cascio coined the term "SEHI" in contrast to another term, "super-empowered angry men," which *New York Times* columnist Thomas Friedman used in his writing about terrorism in a globally networked age. Osama bin Laden, Friedman wrote, seeks to create super-empowered angry men who feel capable of leaving their mark on the world, in terrible ways.[22] In response, Cascio explains:

> The core of the "super-empowered angry individual" (SEAI) argument is that some technologies may enable individuals or small groups to carry out attacks, on infrastructure or people, at a scale that would have required the resources of an army in decades past. . . . But angry people aren't the only ones who could be empowered by these technologies. As a parallel, the core of the "super-empowered hopeful individual" (SEHI) argument is that these technologies may also enable individuals or small groups to carry out socially beneficial actions at a scale that would have required the resources of a large NGO or business in decades past.[23]

SEHIs don't wait around for the world to save itself. They invent and spread their own humanitarian missions. More importantly, they are "able to do so with smaller numbers, greater speed, and a far larger impact" than a slow-moving, risk-averse organization. Of course, in an ideal world, SEHIs would be able to band together and scale up their efforts—to avoid making redundant efforts, to learn from each other's mistakes, to amplify each other's abilities to make a difference. Disorganized SEHIs would have a hard time making significant strides. But organized SEHIs—well, they could change everything.

So a year after the World Without Oil experiment, Cascio and I teamed up at the Institute for the Future to find as many of those millions of SEHIs as possible—to give them a platform for organizing, and a new game to play.

It was called Superstruct, and its promise was simple: Play the game, invent the future.

Superstruct: Inventing the Future of Organization

Every year, the Institute for the Future produces a Ten-Year Forecast. It's a look ahead at the next decade, to identify new economic forces, social practices, and changing environmental realities that will impact the way leading businesses, governments, and nonprofit organizations work, and to define the new challenges they'll face. As we like to say at IFTF, "Ten years is a good, useful horizon—distant enough to expect real changes, close enough to feel within our grasp."[24]

Each Ten-Year Forecast (TYF) has a defining theme, a driving question. In 2008, the TYF program director Kathi Vian decided that the driving question for the next year's forecast would be: What is the future of scale for human organization?

Clearly, we were embarking on a decade of extreme-scale challenges: economic collapse, pandemics, climate change, the continuing risk of global terrorism, and disruptions to our global food supply chain, to name just a few. We knew that existing organizations would have to reinvent themselves in order to simply survive, let alone make a difference.

"We know that the old ways of organizing the human race aren't enough anymore. They're not adapted to the highly connected world we're living in. They're not fast enough, or collaborative enough, or agile enough," Vian wrote during our early brainstorming meetings. "We need to design better ways for the world to work together in the future. We need networked organizations that can solve problems better, move faster, be more responsive, and overcome the old ways of doing and thinking that paralyze us."

So we wanted to find out: How might businesses, governments, and nonprofit organizations team up to make each other more resilient during crisis? How could existing organizations work together to tackle these planetary-scale problems? How should these entities engage the super-empowered individuals who want to be a part of changing the world—and who will go it alone, for better or for worse, if they don't feel engaged?

Our hunch was that surviving the next decade would require entirely new ways of cooperating, coordinating, and creating together. So we wanted to find a new strategic language for talking about revolutionary ways of working together at extreme scales—language that could completely shift our thinking about how to adapt for the coming decade.

We looked at a lot of potential language, but as soon as we hit on the term "superstruct," we knew we'd found it.

> **superstruct** /ˈsüpər͵strəkt/
> verb trans. [L. *superstructus*, p.p. of *superstruere*, to build upon; *super-*, over + *-struere*, to build. See *super-*, and *structure*.]
> To build over or upon another structure; to erect upon a foundation.[25]

"Superstruct" is a term that shows up most often in the fields of engineering and architecture. To superstruct a building is to extend it, to make it more resilient.

Superstructing isn't about just making something bigger. It's about working with an existing foundation and taking it in new directions, to reach beyond present limits. It means creating flexible connections to other structures, to mutually reinforce each other. And superstructing means growing in strategic and inventive ways so that you can create new and more powerful structures that would have been previously unimaginable.

So *superstruct* really seemed to capture the process of extension and reinvention that we wanted to explore in our Ten-Year Forecast. But what would be the best way to investigate a process that didn't quite exist yet?

My graduate studies background is in a social science called "performance studies," in which one of the core research methodologies is to actually *do*, or *perform*, the thing that you're studying. So we decided to build a superstructure.

We decided to superstruct our own Ten-Year Forecasting project by opening it up to the public. We would conduct our primary TYF research as a live, online six-week collaborative experiment—completely open to anyone who wanted to join us.

We called this experiment, naturally, Superstruct, and we framed it as a massively multiplayer forecasting game. We wanted the world to help us forecast the future of organizing at extreme, or epic, scales in order to survive real global threats and solve real planetary-scale problems. And we committed to using whatever collective forecast our players came up with as the foundation for our annual research report and conference the following spring.

The core creative team for the project was made up of program director Kathi Vian, scenario director Jamais Cascio, and myself, the game director. We spent six months working with a team of a dozen additional IFTF researchers and designers to develop the 2019 scenario, research the game topics, create the immersive content, design the gameplay, and build the website.

The game launched on September 22, 2008, with a press release from a fictional organization called the Global Extinction Awareness System. The press release was dated September 22, 2019.

For immediate release:

September 22, 2019

Humans have 23 years to go

Global Extinction Awareness System starts the countdown for Homo sapiens.

Based on the results of a yearlong supercomputer simulation, the Global Extinction Awareness System (GEAS) has reset the "survival horizon" for *Homo sapiens*—the human race—from "indefinite" to 23 years.

"The survival horizon identifies the point in time after which a threatened population is expected to experience a catastrophic collapse," GEAS president Audrey Chen said. "It is the point from which a species is unlikely to recover. By identifying a survival horizon of 2042, GEAS has given human civilization a definite deadline for making substantive changes to planet and practices."

According to Chen, the latest GEAS simulation harnessed over 70 petabytes of environmental, economic, and demographic data, and was cross-validated by ten different probabilistic models.

The GEAS models revealed a potentially terminal combination of five so-called "superthreats," which represent a collision of environmental, economic, and social risks.

"Each superthreat on its own poses a serious challenge to the world's adaptive capacity," said GEAS research director Hernandez Garcia. "Acting together, the five superthreats may irreversibly overwhelm our species' ability to survive."

GEAS notified the United Nations prior to making a public announcement. The spokesperson for United Nations Secretary-General Vaira Vike-Freiberga released the following statement: "We are grateful for GEAS' work, and we treat their latest forecast with seriousness and profound gravity."

GEAS urges concerned citizens, families, corporations, institutions, and governments to talk to each other and begin making plans to deal with the superthreats.

We chose an intentionally provocative starting point for the scenario for several reasons. First, we wanted players to propose awe-inspiring solutions. So we had to pose a scenario that would inspire a sense of awe and wonder—an epic "What if?" What would you do if you woke up one morning to discover that the world's most trusted supercomputer had calculated the entire human species was as endangered as tigers, polar bears, and pandas are today?

Second, we wanted to learn something new, so we had to push our players to imagine previously unthinkable ideas. We aimed to create a forecasting context so far from their ordinary day-to-day concerns that they would feel free to practice extreme creativity and be comfortable pitching "outlier," or unexpected, ideas.

And third, we wanted to give our players a clear goal, a way to measure their success in the game. The GEAS survival horizon gave us the perfect way to do both. We would challenge our players to work together to extend the survival horizon from the year 2042 as far as we could possibly take it. Each year they added to the horizon would represent a significant milestone in the game. (Advances in the survival horizon would be based on an algorithm factoring in the number of active players, how many game missions they completed, and how many achievements they unlocked.)

To ground the game in some specific forecasting topics, we identified five key areas in which players could make a significant impact on our survival horizon. These were the five superthreats, extreme-scale challenges that posed the greatest threat to humanity's survival. But they weren't just threats—they were also opportunities, key areas for coordinated effort and innovation, among organizations and SEHIs alike.

If you wanted to make a difference in our game world of 2019, you had to pick one of these superthreats and start tackling it with the biggest, most surprising ideas you could come up with. These were the five superthreats:

- *Quarantine* covers the global response to declining health and pandemic disease, including the current respiratory distress syndrome (ReDS) crisis. The challenge: How can we protect and improve our global health, especially in the face of pandemics?
- *Ravenous* focuses on the imminent collapse of the global food system, leading to food safety lapses and shortages worldwide. The challenge: How can we feed ourselves in more sustainable and secure ways?
- *Power Struggle* follows the tremendous political and economic upheaval, as well as quality-of-life disruptions, we may suffer as we attempt to move from oil-based societies to solar, wind, and bio-fuel societies. The challenge: How can we reinvent the way we create and consume energy?
- *Outlaw Planet* tracks the efforts to hack, grief, terrorize, or other-

wise exploit the communications, sensor, and data networks we increasingly rely on to run our lives. The challenge: How can we be more secure in a globally networked society?

- *Generation Exile* looks at the difficulties of organizing society and government in the face of one particular challenge: the disappearance of secure habitats for three hundred million refugees and migrants, who have been forced to leave their homes and in many cases their homelands due to climate change, economic disruption, and war. The challenge: How can we govern ourselves and take better care of each other across traditional geopolitical borders?

To help players quickly grasp the details of this complex scenario, for each of these superthreats we created a short video trailer and a series of news headlines describing unfolding events. We also released an online report, set in the year 2019, outlining some of the dilemmas each of these superthreats might provoke, and how they might interact with and magnify each other. In the report, we emphasized a sense of optimism about humanity's ability to overcome the superthreats.

> The human species has a long history of overcoming tremendous obstacles, often coming out stronger than before. Indeed, some anthropologists argue that human intelligence emerged as the consequence of the last major ice age, a period of enormous environmental stress demanding flexibility, foresight, and creativity on the part of the small numbers of early *Homo sapiens*. Historically, those who have prophesied doom for human civilization have been proven wrong, time and again, by the capacity of our species to both adapt to and transform our conditions.
>
> GEAS does not argue or believe that this future is unavoidable. This is perhaps the most important element of our forecast. This is not *fate*. If we act now—and act with intelligence, flexibility,

foresight, and creativity—we can avoid the final threat. We may even come out of this period far stronger than we were before.

Both the report and the trailers ended with the same call to action: Join us to invent the future of the human species. We announced that volunteers were gathering on an online social network site called Superstruct. And we issued a public invitation—on blogs, on Facebook, on e-mail, on Twitter—to join the network. Our core message: Everyone has a part to play in reinventing the way the world works. And in the end, we attracted 8,647 super-empowered hopeful individuals to contribute their best ideas for the future to our super-structing experiment.

But before they tackled the superthreats, our players had an important first mission: invent their future selves.

Like any present-day social network, our 2019 social network asked you to fill out a personal profile. But our profile was different: it focused on *survivability*. What are the specific skills, resources, and communities you can bring to bear on these superthreats? What are you uniquely qualified to contribute to reinventing the world? We encouraged players to have fun imagining their future selves, but we also told them to keep it real. This was essential. *Don't* invent a fictional character, we told them. This is about real play, not role play. We want to know who you *really* think you might be in 2019. Feel free to dream big, but make sure it's grounded in reality.

Here's the profile. How would you answer these questions? Remember: It's not who you are today. It's who you might be in the future.

YOUR WORLD IN 2019

Where do you live?

Who do you live with?

What do you do? Where do you work?

What matters to you most?

How did you get to be this person? Was there a particular
turning point for you in the past ten years?

YOUR UNIQUE STRENGTHS

What do you know more about than most people? Tell us
about your skills and abilities.

Who do you know? Tell us about the communities and groups
you belong to, and what kinds of people are in your social
and professional networks.

This first mission helped immerse players in the future. It required them to vividly imagine the year 2019, and how their work and lives might be different by then. It also helped them identify specific personal resources they could bring to bear on the superthreats. At heart, Superstruct was about figuring out new roles for individuals, organizations, and communities to play in much bigger, longer-term efforts to make life on this planet better. To accomplish this goal, we had to help our players make direct connections between their current skills, resources, and abilities and the demands of the future.

If you were to ask an ordinary person if they could personally fix the economy, stop a pandemic, or prevent famine, they probably wouldn't even know where to begin. So we gave the players a specific place to start: in their own communities, groups, and social networks, using whatever they knew best as a foundation for suggesting solutions.

Finally, this mission gave us some concrete data about our players. We asked players to tell us, confidentially, a little bit about who they were in 2008, to help us put their forecasts and ideas into context. This 2008 data didn't appear on their public profile; it was only to help during the research process, so we could cross-reference their future ideas by real-life age, location, and occupation. We wanted to find out more about how SEHIs think of themselves and what kinds of projects they are most likely to tackle.

So who superstructs? Here's what we found out.

Out of just under nine thousand forecasters who joined the effort, the vast majority were between the ages of twenty and forty, with the rest spread out like a bell curve. Our youngest player was ten (he was particularly interested in the future of food, especially "lab-grown meat," which is in fact an emerging food technology). Our oldest player was ninety (she was interested in the future of education).

We had players from forty-nine out of fifty U.S. states and more than one hundred countries worldwide. We had a startlingly diverse group of professionals, including chief engineers, chief technical officers, chief creative officers; longshoremen, hotel concierges, and museum curators; astrophysicists, atmospheric scientists, mathematicians; nurses, plumbers, and photographers. There were also numerous college and graduate students, senior executives, members of the armed forces, and public servants.

What expertise did these diverse participants bring to the game? Players identified expertise in areas as wide as labor activism, transportation and logistics, and robotics; specialty coffee, the comic book industry, the steel industry; immigration, forestry, and fashion; tourism, health care, and journalism; chemical engineering, caregiving, and e-commerce; consulting, defense, and human resources; forensics, human rights, and nanotechnology.[26]

These were among the skills and resources we asked players to bring to bear on the five superthreats. And we asked them to tackle the superthreats in a very specific way: by inventing superstructures.

⟳ HOW TO INVENT A SUPERSTRUCTURE

This is a game of survival, and we need you to survive.

We're facing superthreats, and we need to adapt.

The existing structures of human civilization just aren't enough. We need a new set of superstructures to rise above, to take humans to the next stage.

You can help. Superstruct now. It's your legacy to the human race.

Q: WHAT'S A SUPERSTRUCTURE?

A: A superstructure is a highly collaborative network that's built on top of existing groups and organizations.

THERE ARE FOUR TRAITS THAT DEFINE A SUPERSTRUCTURE:

1. A superstructure **brings together two or more different communities** that don't already work together.
2. A superstructure is designed **to help solve a big, complex problem** that no single existing organization can solve alone.
3. A superstructure **harnesses the unique resources, skills, and activities of each of its subgroups.** Everyone contributes something different, and together they create a solution.
4. A superstructure is **fundamentally new.** It should sound like an idea that no one's tried before.

Q: WHAT KINDS OF GROUPS CAN COME TOGETHER TO FORM SUPERSTRUCTURES?

A: Any kind of group at all. For profit and not-for-profit, professional and amateur, local and global, religious and secular, online and offline, fun and serious, big and small.

ANY EXISTING COMMUNITY CAN BE ADDED TO A SUPERSTRUCTURE! HERE ARE SOME EXAMPLES:

- Companies
- Families
- People who live in the same building or neighborhood
- Industry and trade organizations
- Nonprofits and NGOs
- Annual conferences or festivals
- Churches

- Local or national governments
- Online communities
- Social network groups
- Fan groups
- Clubs
- Teams

Q: HOW DO I CREATE A NEW SUPERSTRUCTURE?

A: Start by picking a community that you already belong to. What could your community uniquely contribute to solving one or more of the superthreats? And who else do you want to work with to make it happen?

When you're ready to share your idea, create a new wiki article. Use the wiki fields (name, motto, mission, who we need, how we work, and what we can accomplish) to describe your new superstructure.

Q: I'VE MADE A SUPERSTRUCTURE. WHAT NEXT?

A: When you have a basic description of your superstructure in place, invite other SEHIs and your own friends, colleagues, neighbors, and networks to join.

If you've made your superstructure public, keep an eye on your wiki to welcome new members and to see how the super-structure evolves. If you've made your superstructure private, be sure to check back often to approve new members so they can help you build your superstructure.

Together, your Superstruct members can keep editing the wiki until it describes exactly the way you think your super-structure should work.

DON'T STOP NOW!

Once you've created your first superstructure, there's lots more to do. You can create superstructures for other super-threats. Or you can design spin-off superstructures from your

original superstructure. You can invent competing superstructures, or bigger superstructures to swallow up superstructures
that are already existing.

Keep superstructing, and surprise us with your big ideas!

The most important rule for inventing a superstructure was that it should
be unlike any existing organization. It should be a fundamentally new combination of people, skills, and scales of work. But it also had to be a plausible
approach to a problem—a way to give people who don't ordinarily work
on challenges like hunger, pandemic, climate change, economic collapse, or
network security a way to make a difference.

Inventing a superstructure was the core element of gameplay; it was how
players earned survivability points (up to one hundred), which tallied into
a total survivability score. The more thoughtful, clearly explained, creative,
and surprising the superstructure, the more points a player earned. A player
could also earn points by joining and contributing ideas to other players'
superstructures.

What, exactly, is a survivability score? We described it to players as follows:

Your Survivability Score is a number between 0 and 100 that appears in your Survival Profile. When you first join, you have a
score of 0. Any score higher than 0 means you personally are becoming more and more important to the survival of the species. If
you achieve a score of 100, you personally are absolutely central
to the future of the human race.

In other words, our scoring system wasn't meant to be competitive, but
simply to represent your personal progress.

Let's take a look at some brief descriptions of some of the particularly high-
scoring superstructures, and the SEHIs who created them.

WE HAVE THE POWER—ENERGY-HARVESTING CLOTHES

You don't need to buy power from an energy company. You can make your own power. What you wear every day can help you collect and save energy, which you can use to power your laptop, your cell phone, your MP3 player, or to provide heat.

Think: Jackets with solar panels that collect energy and can be used to provide electric heat when you wear the jacket at night. Headbands with solar-paneled flowers that collect the energy you need to power your iPod. Fringed skirts that harness wind energy and store it in a tiny battery that you can detach and use to power anything at all. A belt with a sound wave collector that turns environmental noise into an energy source.

We're creating and collecting designs for all kinds of wearable energy sources. We'll make working prototypes of these designs and present them in a We Have the Power fashion show. We need your help sharing and improving these designs so that as many people as possible can harvest their own energy.

The We Have the Power superstructure was founded by SEHI Solspire, or, in real life, Pauline Sameshima, an assistant professor in the department of teaching and learning at Washington State University. She led her design class in creating a series of real, working prototypes and impromptu campus fashion shows for clothes that incorporated the kinds of wearable energy technologies described above. Their SEHI mission: to use rapid prototyping and design innovation to tackle the Power Struggle superthreat, and help invent the future of energy.

SEEDS ATMS—WITHDRAW YOUR FOOD FOR FREE

Food shouldn't cost anything. Seeds also shouldn't cost anything. That's why this superstructure has been created: to build a

Seeds ATM network, so anyone who needs seeds can easily go to an ATM and get free seeds.

What we want to accomplish: spread the GYO (grow your own) food concept, as well as set the foundations of a bigger free-food network.

We envision a network of secure Seeds ATMs installed at bank locations worldwide. However, as a working prototype, we propose a really simple hack: gumball machines. We will fill them with seeds and set them to not need money, or to simply require a penny. We will install them outside grocery stores and farmers' markets.

This superstructure was invented by SEHI Jorge Guberte, a twenty-five-year-old digital artist in São Paulo, Brazil. With no direct connection to the food industry or agriculture, he proposed a completely unexpected, extreme-scale solution to the Ravenous superthreat. His SEHI mission: to make access to food a basic civic right, and help invent the future of how we feed ourselves.

THE DEMOCRATIC CENTRAL AFRICAN REPUBLIC (DCAR)—THE REFUGEE STATE

DCAR, the Democratic Central African Republic, is a "weakly statelike entity," or WSLE, a 16 million person quasi-state entity in east central Africa in a contested region where the neighboring countries have largely lost the will to continue fighting but will not allow each other to declare victory.

Humanitarian efforts in this disordered area therefore had to provide many basic functions of the state, such as identity services, issuing what amounted to passports, issuing a basic electronic currency, and generally trying to keep life going until somebody asserted governance over the area.

Eventually, after four or five years of interim governance using electronic democracy software and biometric cell phones, DCAR has begun to have semiofficial, quasi-state status. Like Taiwan, it cannot safely assert full sovereignty, but the shells of the previous governments of the region have technically passed their legitimacy to the refugee councils of DCAR, and as long as nobody raises an army, nobody seems to mind self-organizing refugees trying to manage their lives until the governments settle their territorial disputes.

Anybody can support DCAR. You just have to remember how important it is that refugees get political rights to manage their own lives, just as we do. Being a refugee is hard enough without being oppressed too!

If you want to get more involved, display the DCAR flag wherever you can to let people know that DCAR still matters.

The DCAR superstructure was founded by SEHI Hexayurt, or Vinay Gupta, a noted world expert on disaster relief. He is also the inventor of the Hexayurt, an inexpensive, lightweight shelter designed to provide sustainable housing for refugees. He wanted to address the Generation Exile superthreat in the context of the ongoing African refugee crisis. His SEHI mission: to help invent the future of peace and government.

NONE OF THESE ideas will reinvent the way the world works on its own. But alongside the more than five hundred other superstructures that players created, they effectively prove a new reality: that problem solving at extreme scales can involve ordinary people; that all scales of human organization can combine and recombine in startling ways; that continuous reinvention is not only possible, it's an evolutionary imperative for the next decade.

WE RAN SUPERSTRUCT as a live forecasting experiment for six weeks. So what were the final results?

After the game, our players inventoried and organized their efforts into a catalog of solutions called the Whole Superstructure Catalog (a play on Stewart Brand's *Whole Earth Catalog*), which you can view online, at Superstruct. wikia.com.

In addition to their catalog of 550 superstructures, our players created more than a thousand vivid first-person accounts of the superthreats, told in videos and photos, blogs and Twitter updates, Facebook messages and podcasts. This world lives online as a resource for other forecasters, policy makers, educators, and interested individuals to explore and analyze.

We set up traditional discussion forums to provide players with a sounding board for strategies they wanted to apply in their superstructures. The players held court across more than five hundred different forum topics, such as "Networking the offline world: How do we reach out to people who aren't online?"; "What can we do with bicycles: Beyond exercise, how can we use bicycles to help solve some big problems?"; and "Art for art's sake: What is the role of arts in 2019? What role can art play during times of epic crisis?" There's enough reading material on these forums to comprise dozens of future-forecasting reports—and it's all saved online for public browsing.

In terms of gameplay, we had nineteen players achieve a survivability score of 100—the equivalent of winning the staff of life in *Spore*, or creating a level 80 character in *World of Warcraft*. We invited these nineteen players to become our "SEHI 19," and we extended invitations to them to continue collaborating with the Institute for the Future. All of the players, but in particular the SEHI 19, became a kind of superstructure for IFTF itself.

Finally, our Ten-Year Forecast team spent six months analyzing the results of the forecasting game. We prepared the year's TYF research report, "Superstructing the Next Decade," developing the themes explored in our scenario and analyzing the most promising superstructing methods demonstrated by our players. We've since made these materials—including a set of "Superstruct strategy cards," a visual map of the Superstruct ecosystem, and three alternate scenarios for the next fifty years of planet craft—available to the public on the Institute for the Future's website.[27]

Oh, and just in case you're wondering—how much good did our players collectively do, according to the game's Global Extinction Awareness System?

By the end of the six-week game, the players had pushed the survival horizon for the human species to the year 2086—or one year more for every thousand survivability points they collectively earned. That's forty-four more years they earned us on this planet—enough time for two more generations of potential super-empowered hopeful individuals to be born and to start working on these problems with us.

The entire experience is perhaps best summed up by a Twitter post from one of our players. It epitomizes exactly what IFTF hoped to accomplish with the game.

"This is my favorite vision of the future, ever," he wrote. "Because it's the first one I feel personally capable of making a difference in."

Superstruct was designed to wake gamers up to the possibility of making the future together—a critical first step to increasing our collective engagement with global superthreats. But to produce real change in the world, it's not enough to spread a feeling of super-empowerment and hope. We also have to build up actual world-changing capacity among gamers. We have to help them cultivate the specific future-making skills and abilities, and acquire the practical knowledge they need, in order to increase their chances of making a real and sustained difference.

Super-empowered hope and collaborative creativity must be combined with *practical learning* and *real capacity development*. And it can't just happen in the parts of the world where computer and video game technology is already pervasive. World-changing games must be custom designed specifically for the most impoverished regions of the world, where future-making skills are most urgently needed—for example, developing areas in much of Africa.

These are the two key insights that led me to my most recent game, EVOKE.

EVOKE: A Crash Course in Changing the World

EVOKE is designed to empower young people all over the world, especially in Africa, to start actively tackling the world's most urgent problems—poverty, hunger, sustainable energy, clean-water access, natural disaster preparation, human rights.

Dubbed as a "crash course in changing the world" and produced for the World Bank Institute, the learning arm of the World Bank, EVOKE is a social network game designed to help players launch their own world-changing venture in just ten weeks. It's playable on computers, but it's optimized for mobile phones—the most ubiquitous social technology in Africa.

The world of EVOKE is set ten years in the future. The story, told in the form of a graphic novel, follows the adventures of a secret superhero network based in Africa. The network is made up of "stealth social innovators," a concept we invented for the game.

Excerpt from episode eight of EVOKE.
(Jacob Glaser, World Bank Institute, 2010)

Social innovation, of course, is a real concept—and an increasingly important method of tackling poverty worldwide. It means applying entrepreneurial ways of thinking and working to solve social problems that are ordinarily tackled by governments or by relief and aid agencies. The key principle of social innovation is that anyone, anywhere, can start their own project or business venture to try to solve a social problem. Also referred to as "social entrepreneurship," it emphasizes taking risks, understanding the local context, and looking for breakthrough innovations, rather than applying standard, cookie-cutter solutions.

So what is *stealth* social innovation? In the world of EVOKE, social innovators tackle social problems with superheroic secrecy and spectacle—public and yet mysterious, like Batman or Spider-Man—in order to capture global imagination so that the solutions have a real chance to catch on and spread virally. EVOKE superheroes are particularly known for applying an innovation method referred to by real development experts today as "African ingenuity."

Erik Hersman, a technologist and editor for the blog AfriGadget, is a leading proponent of African ingenuity. Hersman, who grew up in Sudan and now lives in Kenya, describes it as follows:

> A Malawian boy creates a windmill from old bicycle parts and sheet metal. A Kenyan man fabricates welding machines from scrap metal, wood and copper wire. An Ethiopian entrepreneur makes coffee machines from old mortar shells. A Malawian scientist invents a new micro–power plant that uses sugar and yeast. A South African youth makes a working paraglider from plastic bags, rope and bailing wire. Though you might not hear those stories in the international press, these are just a few of the incredible tales of African ingenuity happening every day in thousands of villages, godowns, industrial areas, roadside shops and homes throughout the continent. Africans are bending the little they have to their will, using creativity to overcome life's challenges.[28]

Many experts on Africa, including Hersman, believe that the people who tackle the hardest problems in the developing world today will be the ones most capable of solving any crisis, anywhere in the world, in the future. Indeed, problem solvers in Africa today may leapfrog past the rest of the world, coming up with cheaper, more efficient, and more sustainable solutions, simply because they have no other choice. The obstacles they face are so enormous, and the resources they have so limited, that their solutions *must* be more creative, more resourceful, and more resilient than traditional solutions developed by the rest of the world.

EVOKE is designed to help players become a part of the emerging culture of African ingenuity—to build up their social innovation skills today so they have a real chance to become the world's superheroes in the future.

So how does the gameplay work? Over the course of a ten-week "season," players are challenged to complete a series of ten missions and ten quests. Each week's challenges are focused around a new "urgent evoke."

An **evoke**, in this game world, is an *urgent call to innovation*, an electronic SOS message sent from a city in crisis to the secret problem-solving network in Africa. In the first two episodes of the game, for example, the EVOKE network is called upon to help prevent a famine in Tokyo and to rebuild following a collapse of the energy infrastructure in Rio de Janeiro.

After reading the urgent evoke online, players are challenged to respond in the real world—and get real, firsthand experience tackling an urgent crisis on a small and local scale. Consider the first two EVOKE missions.

URGENT EVOKE: Food Security

More than a billion people go hungry every day. This week, YOU have the power to change at least one of those lives. Your objective: Increase the food security of at least one person in your community. Remember: Food security isn't about providing temporary help or a single meal. It's about long-term solutions to hunger and food shortages. Here are some ideas to get you started:

- Help someone start a home garden.
- Volunteer at a local community garden.
- Invent a way to make it easier for people in your community to share the food they have with others.
- Create a resource for local farmers.

URGENT EVOKE: Power Shift

Today, less than 10 percent of global electricity is produced by sustainable energy sources. This week, discover YOUR power to help change that number. Your objective: Design a new way to power something you use every day. Take a look around you. Something YOU use or do every day could be powered differently—with solar power, wind power, or kinetic power, for example. Maybe it's your mobile phone. Maybe it's the light you use to read at night. Your solution should be cheaper or more sustainable than your current power source.

To help them brainstorm creative solutions to these challenging tasks, we provide players with secret "investigation files" that document social innovations already happening in Africa and other parts of the world—projects that can spark their own African ingenuity and inspire their own efforts.

In order to receive credit for their missions, players must share a blog post, video, or photo essay documenting the effort they made and what they learned. Other players review the mission evidence to verify it and to award EVOKE powers: plus-one spark, for example, or plus-one knowledge sharing, or plus-one local insight. Through the course of the game, by completing all ten missions, players build up a personal portfolio of world-changing efforts (their collection of blog posts, videos, and photos), as well as a profile of their unique future-making attributes (an interactive display of all the EVOKE powers they've earned).

Meanwhile, players are also challenged to complete a series of ten online quests. These personal quests are designed to help players discover their own unique "origin story." The game instructions explain, "In comic books, the origin story reveals how a character became a superhero—where their powers came from, who inspired them, and what events set them on a path to change the world. Before YOU can change the world, you need to figure out your superhero origins." Over time the players' quest log becomes a kind of world-changing calling card, describing, for example, their secret identity. The quest log would include answers to questions like "What are three things you know more about, or do better, than most of your friends and family?" and "What three personality traits or abilities make you stand out from the crowd?" The quest log also represents their heroic call to action when they answer questions like "If you had the power to convince today one person—or a hundred people, or a million people—to do one thing, who would it be, and what would you call on him or her to do?" By completing these introspective quests, players aren't just learning about their own strengths or charting their own future.

They're also developing the foundations for a multimedia business plan that they can use to attract collaborators, mentors, and investors.

Robert Hawkins, a senior education specialist at the World Bank Institute, first came up with the idea for a social innovation game. "The demand is so great for a game like this," Hawkins told me when he first invited me to join the project as its creative director. "We keep hearing from African universities that they need better tools to engage students in real-world problems and to develop their capacities for creativity, innovation, and entrepreneurial action. This game needs to be a response to that desire, to serve as an engine for job creation now and in the future." In fact, the game was promoted to university students across English-speaking Africa as "Free job training—for the job of inventing the future."

Not only are EVOKE players learning real-world skills, they're also earning real-world honors and rewards. Players who successfully complete ten online missions in ten weeks receive a special résumé-worthy distinction: official certification as a World Bank Institute Social Innovator. Top players also earn postgame mentorships with experienced social innovators, and scholarships are awarded so they can share their vision for the future at the annual EVOKE Summit in Washington, D.C.

In the first trial of the game, run in the spring of 2010, we enrolled more than 19,000 young people from over 150 countries, including more than 2,500 students from sub-Saharan Africa—making it the largest collaborative online problem-solving community in Africa to date.

Collectively, in just ten weeks, this founding group of players completed more than 35,000 future-making missions together, documented on the EVOKE network. More importantly, as their final challenge, they proposed more than a hundred new social ventures—creative enterprises they planned to undertake in the real world, with the support of seed funding and ongoing mentorships from the World Bank Institute. These EVOKE-inspired ventures include:

- Evokation Station, a pilot program created and managed by high school students in Cape Town, South Africa, and designed to give people the skills and knowledge to grow their own food for their families and as a source of income. The program is currently being tested in one of the poorest communities in Cape Town, Monwabisi Park, an informal settlement, or squatter camp, of more than twenty thousand displaced people who have been living for twelve years without running water, sanitation, proper houses, roads, or access to health care and employment.

- Solar Boats, a project by and for young women in Jordan, with the goal of converting more than 120 glass boats in the Gulf of Aqaba to solar-powered ones, in order to save on fuel, decrease pollution of the Red Sea coral and sea life, and lead to cleaner beaches and lower-cost boating.

- Spark Library, a venture developed by a U.S. graduate student in architecture, to design and pilot a new kind of crowdsourced library across sub-Saharan Africa. In order to check out a book from a Spark Library, you must first contribute a piece of local or personal knowledge, in order to help build up a database of indigenous or traditional knowledge about the environment, cultural practices, and natural resources.

As I write this chapter, plans to develop future seasons of EVOKE are already under way, based on its early success. New seasons of the game will focus attention and engagement on a single issue, such as energy, food security, or women's rights. Meanwhile, the first season of the game—EVOKE's core curriculum—will be translated into Arabic for the Middle East, Spanish for Latin America, Mandarin for China, and more, in order to reach even more students. And in order to support EVOKE play in regions of Africa without reliable Internet access, episodes from the first season of EVOKE are being compiled into a single graphic novel, with all of the missions and quests adapted into workbook-style exercises. SMS-based interactivity—as most young people in the developing world do have access to mobile phones—will ensure that these "pen and paper" players are still connected to the global EVOKE network.

The goal of all of these adaptations? To ensure, over time, that every young person on this planet receives an education in urgent problem solving and planet crafting—and has free and open access to a global network of potential world-changing collaborators, investors, and mentors.

SO HOW MIGHT future-making games like World Without Oil, Superstruct, and EVOKE evolve in a best-case-scenario future?

At the end of Superstruct, all of the IFTF game masters had an opportunity to select and honor their favorite superstructure during an online streaming Superstruct Honors broadcast. I chose a superstructure called The Long Game, proposed by player Ubik2019, one of Superstruct's most active players. The Long Game represents, to me, what future-making games must aspire to become by the end of the twenty-first century: an *epic collaboratory* for our most awe-inspiring global development efforts.

In real life, Ubik2019 is Gene Becker, formerly the worldwide director of product development for extreme performance and mobility at Hewlett-Packard, and now the founder and managing director of Lightning Laboratories, an emerging-technology consulting company that works with a range of Global 2000 companies and preinvestment start-ups. Becker brought to Superstruct a particularly keen sensibility about how to develop initiatives on a global scale, and how to leverage new network technologies for innovation. Here is Becker's best idea for a new superstructure:

THE LONG GAME

Fostering a long-term mind-set by playing a game that lasts a thousand years.

> *Who we need:* SEHIs who believe that a long-term mind-set and a playful approach to life can help us to become a better people, make better choices about our actions and their consequences,

potentially avoid the kind of supercrises we are facing here in 2019, and give every person on the planet the opportunity to create a meaningful legacy to the human race.

What we can accomplish: If you put just one dollar into an investment today that has an average real return of 3 percent per year after inflation and taxes, in a thousand years it would be worth $7 trillion. Now think about what your descendents thirty generations in the future might be able to do with such capital—and think about how you might communicate your wishes to them about how they would spend it.

Now consider how we might invest our *nonfinancial* capital—intellectual, natural, social, familial, genetic—in such a way that it compounded its value over time. Such a rich gift we could endow for the future of humanity . . .[29]

A thousand-year game, combining financial and nonfinancial investment of our most important resources—how exactly would such a game work?

During the Superstruct experiment, we brainstormed different ideas, focusing largely on structure rather than theme, story, or content. For example, we imagined the entire world setting aside one day each year to play the game, as a kind of global holiday. Of course, like all good games, participation would be optional. But the supergoal of the game would be, by the end of one thousand years, to engage virtually 100 percent of the human population in playing.

Enthusiastic players could spend as much time as they wanted throughout the year preparing for the global game day. Casual players could simply show up to a game site (online or in the real world) and take part for a few minutes, a few hours, or even all twenty-four hours in the year's game day.

That entire global game day would represent one "move" in the game. And perhaps, we imagined, The Long Game would be played in rounds of fifty moves each. So if you played The Long Game your entire life, you would hope to be able to experience a complete round at least once, if not twice.

Every tenth move would represent a bigger and more significant occasion, to provide a kind of momentous leveling-up occasion each decade. Each twenty-fifth and each fiftieth move, the halfway mark and the end of each round, would be an even more momentous occasion—each time the culmination of a quarter century of gamers' efforts.

What specifically would making a move in the game entail? We envisioned a combination of events. Social rituals and circle games to build common ground. Crowdsourced challenges and collective feats—in the style of a traditional barn raising—to focus the world's energy and attention on a single problem and a single transformation. And forecasting exercises to create shared momentum for the future, and to collectively decide the challenges and themes of the next year's set of games.

No one would ever live to see both the start and the end of the game, of course—not even close. But the game would be a *throughline* for humanity, a tangible connection between our actions today and the world our descendants inherit tomorrow. It would create a sense of awe and wonder, inspiring us to imagine how this massively scaled adventure we are a part of could play out, and to make as meaningful an impact in the game as possible, so we can make a difference in our lifetime that lasts for many lifetimes more.

It's not that hard to imagine people spending their entire lives playing a single game. Many *World of Warcraft* gamers have now been playing their favorite game for nearly an entire decade already; so has the *Halo* community. Countless among us spend a lifetime mastering a game like chess, poker, or golf.

And we already have a historical precedent for societies successfully keeping a game tradition alive for an entire millennium—the ancient Greeks ran their Olympiad every four years without interruption for roughly one thousand years.

The Long Game doesn't exist yet. But it just might be what the world needs now.

Aspiring to engage every single human being on the planet in a single game isn't an arbitrary goal. To accomplish that goal would mean transforming the planet and global society in key ways.

It would require every single village in the world to have some level of access to the Internet, via personal computers or mobile phones, so that truly everyone could contribute to the game. Universal Internet access is in its own right a significant and worthy goal. Today, roughly one in four people on the planet has reliable, daily Internet access.[30] When every family in the remote villages of Africa, or in what today are the slums of India, or throughout Nicaragua—when they and everyone else in the world has access to The Long Game, that will mean greater access to education, culture, and economic opportunity as well.

Furthermore, for every person on the planet to play the same game, there would need to be free communications across all geopolitical borders. What would it take before every citizen of North Korea, for example, could play The Long Game?

The Long Game, if we have the will to design it, and if we create the means for universal participation, could be the good game that humanity plays to collectively take us to the next level, achieving a new scale of cooperation, coordination, and cocreation. As Kathi Vian urged in her introduction to the Superstruct Ten-Year Forecast:

> Zoom out. Look at the coming decade from the perspective of millennia of change. Focus on the progress of the universe from the breakthrough structures of the atom to the living cell, the biota, the community of nations, the global economy. This is how the future will be new, by continuing the incredible experiment of reorganization for greater complexity, by creating the next astonishing structural forms in this long evolutionary path.

It seems clear to me that games are the most likely candidate to serve as the next great breakthrough structure for life on earth.

There's no guarantee, of course, that evolution will continue along any given path, other than the path of improved survivability in a given environment. But all of the historic evidence seems to suggest that collaboration improves human survivability, and will continue to do so, as long as we can innovate new ways of working together.

First humans invented language. Then farming, and cities; trade and democratic forms of government and the Internet—all ways of supporting human life and collaboration at bigger and more complex scales.

We have been playing good games for nearly as long as we have been human. It is now time to play them on extreme scales.

Together, we can tackle what may be the most worthwhile, most epic obstacle of all: a whole-planetary mission, to use games to raise global quality of life, to prepare ourselves for the future, and to sustain our earth for the next millennium and beyond.

Reality Is Better

If I'm going to be happy anywhere,
Or achieve greatness anywhere,
Or learn true secrets anywhere,
Or save the world anywhere,
Or feel strongly anywhere,
Or help people anywhere,
I may as well do it in reality.

— FUTURIST ELIEZER YUDKOWSKY[1]

We can play any games we want.

We can create any future we can imagine.

That is the big idea we started with, fourteen chapters ago, as we set off to investigate why good games make us better, and how they can help us change the world.

Along the way, we've gleaned industry secrets—more than thirty years' worth—from some of the most successful computer and video game developers in the world. We've compared these secrets alongside the most important scientific findings of the past decade, from the field of positive-psychology research. We've identified key innovations in the emerging landscape of alternate reality design. And we've tracked how game design is creating new ways for us to work together at extreme scales, and to solve bigger real-world problems.

We have thoroughly assessed all the ways that games optimize human experience, how they help us do amazing things together, and why they enable lasting engagement. As a result, we're now equipped with fourteen ways to fix reality—fourteen ways we can use games to be happier in our everyday lives, to stay better connected to people we care about, to feel more rewarded for making our best effort, and to discover new ways to make a difference in the real world.

We've learned that a good game is simply an unnecessary obstacle—and that unnecessary obstacles increase self-motivation, provoke interest and creativity, and help us work at the very edge of our abilities (*Fix #1: Tackle unnecessary obstacles*).

We've learned that gameplay is the direct emotional opposite of depression: it's an invigorating rush of activity, combined with an optimistic sense of our own capability (*Fix #2: Activate extreme positive emotions*). That's why games can put us in a positive mood when everything else fails—when we're angry, when we're bored, when we're anxious, when we're lonely, when we're hopeless, or when we're aimless.

We've discovered how game designers help us achieve a state of blissful productivity: with clear, actionable goals and vivid results (*Fix #3: Do more satisfying work*). We've seen how games make failure fun and train us to focus our time and energy on truly attainable goals (*Fix #4: Find better hope of success*). We've seen how they build up our social stamina and provoke us to act in ways that make us more likeable (*Fix #5: Strengthen your social connectivity*), and how they make our hardest efforts feel truly meaningful, by putting them in a much bigger context (*Fix #6: Immerse yourself in epic scale*).

If we want to keep learning about how to improve our real quality of life, we need to continue mining the commercial game industry for these kinds of insights. The industry has consistently proven itself, and it will continue to be, our single best research laboratory for discovering new ways to reliably and efficiently engineer optimal human happiness.

———

WE'VE ALSO EXPLORED how alternate reality games are reinventing our real-life experience of everything from commercial flying to public education, from health care to housework, from our fitness routines to our social lives.

We've seen how these games can help us enjoy our real lives more, instead of feeling like we want to escape from them (*Fix #7: Participate wholeheartedly wherever, whenever we can*). We've considered how points, levels, and achievements can motivate us to get through the toughest situations and inspire us to work harder to excel at things we already love (*Fix #8: Seek meaningful rewards for making a better effort*). We've looked at how games can be a springboard for community and build our capacity for social participation, connecting us in spaces as diverse as museums, senior centers, and busy city sidewalks (*Fix #9: Have more fun with strangers*). We've even looked at ways that big crowd games can make it easier for us to adopt scientific advice for living a good life—to think about death every day, for example, or to dance more (*Fix #10: Invent and adopt new happiness hacks*).

These early alternate realities may not represent full, complete, or scalable solutions to the problems they're attempting to solve. But they're vivid demonstrations of what's just now becoming possible. And as more and more of the world's leading organizations and most promising start-up companies begin to test the alternate reality waters, this experimental design space will become an increasingly important wellspring of both technological and social innovation.

FINALLY, WE'VE EXPLORED how playing very big games can help save the real world—by helping to generate more participation bandwidth for our most important collective efforts.

We've looked at crowdsourcing games that successfully engage tens of thousands of players in tackling real-world problems for free—from curing cancer to investigating political scandals (*Fix #11: Contribute to a sustainable engagement economy*).

We've looked at social participation games that help players save real lives and grant real wishes, by creating real-world volunteer tasks that feel as heroic, as satisfying, and as readily achievable as online game quests (*Fix #12: Seek out more epic wins*).

We've learned that young people are spending more and more time playing computer and video games—on average, ten thousand hours by the time they turn twenty-one. And we've learned that these ten thousand hours are just enough time to become extraordinary at the one thing all games make us good at: cooperating, coordinating, and creating something new together (*Fix #13: Spend ten thousand hours collaborating*).

And we've seen how forecasting games can turn ordinary people into super-empowered hopeful individuals—by training us to take a longer view, to practice ecosystems thinking, and to pilot massively multiple strategies for solving planetary-scale problems (*Fix #14: Develop massively multiplayer foresight*).

Very big games represent the future of collaboration. They are, quite simply, the best hope we have for solving the most complex problems of our time. They are giving more people than ever before in human history the opportunity to do work that really matters, and to participate directly in changing the whole world.

ALONG THE WAY to crafting these fourteen fixes, we've inventoried fourteen ways that, compared with our very best games, reality is broken.

Reality is too easy. Reality is depressing. It's unproductive, and hopeless. It's disconnected, and trivial. It's hard to get into. It's pointless, unrewarding, lonely, and isolating. It's hard to swallow. It's unsustainable. It's unambitious. It's disorganized and divided. It's stuck in the present.

Reality is all of these things. But in at least one crucially important way, reality is also *better*: reality is our destiny.

We are hardwired to care about reality—with every cell in our bodies and every neuron in our brains. We are the result of five million years' worth of genetic adaptations, each and every one designed to help us survive our natural environment and thrive in our real, physical world.[2]

That's why our single most urgent mission in life—the mission of every human being on the planet—is to engage with reality, as fully and as deeply as we can, every waking moment of our lives.

That doesn't mean we can't play games.

It simply means that we have to stop thinking of games as only escapist entertainment.

So how should we think of games, if not as escapist entertainment?

We should think of them the same way the ancient Lydians did.

Let's turn back one more time to the provocative history that Herodotus told of why the ancient Lydians invented dice games: so that they could band together to survive an eighteen-year famine, by playing dice games on alternate days and eating on the others.

There are three key values we share in common with the ancient Lydians when it comes to how and why we play games today.

For the starving and suffering Lydians, games were a way to raise real quality of life. This was their primary function: to provide real positive emotions, real positive experiences, and real social connections during a difficult time.

This is still the primary function of games for us today. They serve to make our real lives better. And they serve this purpose beautifully, better than any other tool we have. No one is immune to boredom or anxiety, loneliness or depression. Games solve these problems, quickly, cheaply, and dramatically.

Life is hard, and games make it better.

ORGANIZING LARGE GROUPS of people is also hard—and games make it easier.

Dice games provided the Lydians with new rules of engagement. The rules of engagement were simple: play on these days, eat on those days. But these two simple rules, at least as Herodotus imagined it, supported the Lydians' kingdom-wide efforts to coordinate scarce resources and to cooperate together for the entire duration of the famine—eighteen long years.

It was the institution of daily gameplay that united the kingdom and made it possible to put in so strong an effort over such a long period of time. Increasingly, we, too, are using games to create better rules of engagement and to broaden our circle of cooperation. More and more, we recognize the unrivaled power of gameplay to create common ground, to concentrate our collective attention, and to inspire long-term efforts.

Games are a way of creating new civic and social infrastructure. They are the scaffold for coordinated effort. And we can apply that effort toward any kind of change we want to make, in any community, anywhere in the world.

Games help us work together to achieve massively more.

AND FINALLY, as the Lydians were so quick to realize, games do not rely on scarce or finite resources.

We can play games endlessly, no matter how limited our resources.

Moreover, when we play games, we consume less.

This is perhaps the most overlooked lesson of the story that Herodotus told. For the ancient Lydians, games were actually a way to introduce and support a more sustainable way of life. It was impossible for them to consume their natural resources at the old rate, so new games enabled them to adopt more sustainable habits.

We are just starting to realize this possibility for ourselves today. We are starting to question material wealth as a source of authentic happiness. We are starting to look for ways to avoid exhausting the planet, and each other, with our escalating need for more stuff. We are looking to increase our wealth of experiences, relationships, and positive emotions instead.

The closer we pay attention to the real and completely renewable rewards we get from games, the better we understand: games are a sustainable way of life.

WE SHARE with the ancient Lydians these three timeless truths about games: Good games can play an important role in improving our real quality of life. They support social cooperation and civic participation at very big scales. And they help us lead more sustainable lives and become a more resilient species.

But we are also different from the ancient Lydians, in one crucial way, when it comes to the games we play.

Their dice games did many things, but what they did *not* do, as far as we know from Herodotus, is actually solve the problem of famine itself. The games eased the problem of individual suffering. They solved the problem of social disorganization. They solved the problem of how to consume fewer scarce resources. But they did not solve the problem of the collapse of the food supply itself. They did not bring the greatest minds together to test and develop new ways of getting or making food.

Today, games have sufficiently evolved to support this fourth crucial function. Games today often have content—serious content—that directs our attention to real and urgent problems at hand. We are wrapping real problems inside of games: scientific problems, social problems, economic problems, environmental problems. And through our games, we are inventing new solutions to some of our most pressing human challenges.

The ancient Lydians just had dice games. Today, we are developing a much more powerful kind of game. We are making *world-changing* games, in order to solve real problems and drive real collective action.

SO WHAT ever happened to the ancient Lydians?

If Herodotus is to be believed, their story has a surprise happy ending.

After eighteen years of dice games, Herodotus writes, the Lydians saw that there still was no end to the famine in sight. They realized that they couldn't simply survive the famine by waiting it out and distracting themselves from their misery. They had to rise to the occasion and tackle the obstacle directly.

And so it was decided: they would play one final game together.

The kingdom's population was divided into two, Herodotus tells us, and by the chance drawing of lots, it was decided which half of the population would stay in Lydia and which half would set out in search of more hospitable land.

This final game is what led the Lydians to their own epic win—an unexpected but profoundly triumphant solution to the problem of the famine. The food resources of Lydia, it turned out, could much more easily sustain half as many people, and indeed we know from other historical accounts that the kingdom subsequently not only survived for centuries more, but flourished. Meanwhile, according to Herodotus, the Lydians who'd sailed off in search of a new home settled to great success in what is now the Tuscan region of Italy, where they developed into the highly sophisticated Etruscan culture.

The Etruscans, of course, are known today as the single most important influence on Roman culture. Historians widely agree that it was the Etruscans who originally developed the great skills of urban planning and civil engineering, and that it was the Etruscans' efforts to advance art, agriculture, and government that provided the foundations for the world-changing Roman Empire—and, therefore, much of Western civilization as we know it.

But were the game-playing Lydians really so influential in the course of human civilization? Competing histories of the Italian region have claimed for centuries, as a point of local pride, that the Etruscans were native to the region, not immigrants. Meanwhile, like many of the histories written by Herodotus, this account of the Etruscans' origins has been met with some skepticism. The tale of the starving Lydians and their gaming is so fanciful that many modern historians have dismissed it as a myth or fable, perhaps inspired by facts but not bound by them.

However, recent scientific research appears at long last to conclusively confirm several key details of Herodotus' account of the Lydians, both of the famine they faced and their eventual mass migration.

Geologists today believe that a catastrophic global cooling occurred between the years 1159 and 1140 BC—a nineteen-year time frame they've identified using tree-ring dating.[3] A tree ring is a layer of wood produced during one tree's growing season; during droughts and famines, tree rings are extremely narrow compared with normal seasons. By examining the rings in petrified trees, geologists have concluded that global cooling caused severe droughts and famines lasting for almost two decades in the twelfth century BC, particularly in Europe and Asia. Historians now believe this global cooling may have prompted the eighteen-year famine in Lydia that Herodotus described.

Meanwhile, in 2007, a team led by Alberto Piazza, a geneticist at the University of Turin, Italy, made what was widely considered a breakthrough find-ing in human genetics. The research team analyzed the DNA of three different present-day Tuscan populations known to be direct descendants of the Etruscans. They discovered that the Etruscans' DNA was much more closely linked with near-Eastern peoples than with other Italians, and, cru-cially, they found one genetic variant that is shared only by people from Tur-key, the region once populated by the Lydians. As Piazza reported at the time of his team's discovery, "We think that our research provides convincing proof that Herodotus was right, and that the Etruscans did indeed arrive from an-cient Lydia."[4]

With this modern-day scientific confirmation of two crucial details of Herodotus' account, the legend of the ancient Lydians takes on new signifi-cance. An astonishing claim becomes suddenly much more plausible: we may owe much of Western civilization as we know it to the Lydians' ability to come together and play a good game.

It turns out the dice games weren't just a way to be happier during difficult times. They were also teaching the entire society to work together wholeheart-edly toward collectively agreed-upon goals. They were training the Lydians to hold on to a sense of urgent optimism even in the face of daunting odds. They were building a strong social fabric. And they constantly reminded every Lyd-ian that they were a part of something bigger.

These are exactly the good game skills and abilities that the ancient Lydians drew upon in order to survive catastrophic climate change and reinvent their own civilization.

If they did it then, we can do it again today.

We have been playing computer games together for more than three decades now. By that count, we've accumulated our own eighteen years' worth of pre-paratory good gaming, and then some. We have the collaboration superpowers. We have the interactive technology and global communication networks. We have the human resources—more than half a billion gamers and counting.

More than three thousand years after the ancient Lydians harnessed their game skills and abilities to reinvent the world, we are ready to do the same.

We are ready for humanity's next epic win.

———

WE CAN no longer afford to view games as separate from our real lives and our real work. It is not only a waste of the potential of games to do real good—it is simply untrue.

Games don't distract us from our real lives. They *fill* our real lives: with positive emotions, positive activity, positive experiences, and positive strengths.

Games aren't leading us to the downfall of human civilization. They're leading us to its reinvention.

The great challenge for us today, and for the remainder of the century, is to integrate games more closely into our everyday lives, and to embrace them as a platform for collaborating on our most important planetary efforts.

If we commit to harnessing the power of games for real happiness and real change, then a better reality is more than possible—it is likely. And in that case, our future together will be quite extraordinary.

Acknowledgments

I could not have written this book without the inspiration, collaboration, advice, and support of the following individuals. I wholeheartedly thank:

Chris Parris-Lamb, my agent at the Gernert Company, for having the vision for this book before I did, for convincing me to write it, and, most importantly, for donning a blindfold and running through a labyrinth the very first time we met. **+50 Courage**

Laura Stickney and Alex Bowler, my wildly talented editors at Penguin Press and Jonathan Cape, for so much editorial wisdom and encouragement. Thank you for finding what was important in the book and bringing real clarity to my ideas. (And for ensuring that I used the word "awesome" *fewer* than a hundred times in the final manuscript.) **+50 Epic Guidance**

Everyone at Penguin Press, Jonathan Cape, and the Gernert Company, for taking seriously the idea that games can make us better and change the world—and for lending your great skills and talents to this project. **+100 Teamwork**

My brilliant colleagues at the Institute for the Future, for always bringing the next decade into focus, and especially my mentors, who taught me how to think about the future: Marina Gorbis, Jean Hagan, Bob Johansen, and Kathi Vian. **+100 Foresight**

Everyone at the Leigh Bureau, for helping me hone my story about the power of games, and for finding opportunities to deliver it to amazing communities and organizations around the world. **+50 Encouragement**

The conference organizers who invited me to give the talks that inspired this book—Hugh Forrest (for SXSW), Eric Zimmerman (for the Game Developers rant), Susan Gold (for the IGDA Education Summit), and June Cohen, Kelly Stoetzel, and Chris Anderson (at TED); and the Conference Associate program at the Game Developers Conference, for providing aspiring game developers with a foot in the door. **+20 Life Changer**

The great game designers and game researchers whose work inspired and informed this book—most influentially, Edward Castronova, Katherine Isbister, Raph Koster, Frank Lantz, Nicole Lazzaro, and Katie Salen. **+20 Big Ideas**

The positive psychologists whose research helped me understand why we love games so much; their research is the engine for my game design, especially: Mihály Csíkszentmihályi, Dacher Keltner, Sonja Lyubomirsky, and Martin Seligman. **+25 Well-being, +25 Happiness, +25 Life Satisfaction, and +25 Flow**

The amazing game developers and creators who taught me the craft—especially Elan Lee, Sean Stewart, Jim Stewartson, Ian Fraser, and Finnegan Kelly. **+100 Creative Genius**

The players of EVOKE, The Lost Ring, Superstruct, CryptoZoo, World Without Oil, Top Secret Dance-Off, Bounce, Cruel 2 B Kind, and Tombstone Hold 'Em, for daring to go where no gamers had gone before. **+50 Gamefulness**

My closest collaborators on these projects—Kiyash Monsef, Robert Hawkins, and Nathan Verrill on EVOKE; Jamais Cascio and Kathi Vian on Superstruct; Ken Eklund and Cathy Fischer on World Without Oil; Greg Niemeyer and Ken Goldberg on Bounce; Ian Bogost on Cruel 2 B Kind; Julie Channing, Edwin Veelo, Toria Emery, and all the global puppet masters on The Lost Ring; and Elan Lee on Tombstone Hold 'Em. **+100 Superheroic Collaboration**

Mike and Paula Monsef, for encouraging me to write and always wholeheartedly participating in my games. **+200 Nurturing**

My parents, Kevin and Judy, who bought a used Commodore 64 for me and my sister when we were in fifth grade so that we could practice writing on Bank Street Writer and learn to program our own games in BASIC. **+500 Love**

My twin sister, Kelly, for her support through the process of writing this book—and for telling me more than ten years ago (in a flash of empathic insight) that, based on my childhood talents and strengths, I should invent a career for myself combining game design and public speaking. It sounded crazy. But she was right. **+1000 Compassionate Willpower**

And most of all, my husband, Kiyash, who is my all-time biggest epic win, and the best possible ally in making a life worth living. "You know, sweetheart, if there's one thing I've learned, it's this: nobody knows what's going to happen at the end of the line, so you might as well enjoy the trip." **+1000 Curiosity, +1000 Wonder, +5000 Meaning**

Appendix 1

HOW TO PLAY

This list is designed to help you learn more about the games in this book—and to get firsthand experience playing them. If you want to become actively involved in the community of people who are already making and playing world-changing games, these resources will show you where to start.

HOW TO FIND OUT MORE

To read more case studies and learn about new and upcoming alternate reality games, forecasting games, happiness hacks, crowd games, and collaboratories, visit the website for this book, **www.realityisbroken.org**.

WHERE TO GET INVOLVED

If you want to help create, playtest, sponsor, or commission a game designed to have a positive impact—to improve players' lives, to solve real problems, or to change the world—join the social network Gameful, at **www.gameful.org**. Other organizations dedicated to a similar mission include Games for Change (www.gamesforchange.org), Games Beyond Entertainment (www.gamesbeyondentertainment.com) and the annual academic Games, Learning, and Society Conference (www.glsconference.org).

WHAT TO PLAY

Many of the alternate reality and world-changing games described in this book are free and available to play online or on your mobile phone. Others are no longer playable, but have been archived online for public viewing. The best online resources for learning about or playing these games are described below. Because many of the games in this book are, at the time of publication, still in beta or prototype form, their availability may change; we will track their availability and the emergence of new games at the book's website, www .realityisbroken.org.

The games below are arranged in alphabetical order, with the chapter in which they are described listed after their name.

BOUNCE

(Chapter 9) A beta version of this cross-generation conversation game, developed at the UC Berkeley Center for New Media by Irene Chien, Ken Goldberg, Jane McGonigal, Greg Niemeyer, and Jeff Tang, is available in English and Spanish at http://heidegger.ieor.berkeley.edu/bounce/.

CHORE WARS

(Chapter 7) A beta version of the chores-management game, created by Kevan Davis, is playable at www.chorewars.com.

COME OUT & PLAY FESTIVAL

(Chapter 9) Find out when and where the annual street festival for new mobile, social games is happening at www.comeoutandplay.org.

THE COMFORT OF STRANGERS

(Chapter 9) Find out more about how to play this social street game, invented by Simon Evans and Simon Johnson, at http://swarmtoolkit.net, or watch a short documentary at http://vimeo.com/1204230.

CRUEL 2 B KIND

(Chapter 10) A short documentary of the game of benevolent assassination, created by Jane McGonigal and Ian Bogost, is available at www.cruelgame

.com, where you can also download a kit for running your own Cruel 2 B Kind game.

DAY IN THE CLOUD

(Chapter 8) You can play the archived version of this in-flight game, developed by Google Apps and Virgin America, wherever you are — even if you're not on an airplane! — at www.dayinthecloud.com.

EVOKE

(Chapter 14) You can join the EVOKE game network for social innovation, created by Jane McGonigal and Kiyash Monsef, and developed by the World Bank Institute and Natron Baxter Applied Gaming, at www.urgentevoke.com.

THE EXTRAORDINARIES (NOW KNOWN AS SPARKED)

(Chapter 12) Join the microvolunteering game, created by Jacob Colker and Ben Rigby, or design your own nonprofit mission, at www.sparked.com. Find out more at http://blog.beextra.org.

FOLD IT!

(Chapter 11) You can solve protein-folding puzzles for science at http://fold .it/portal, a collaboration between the University of Washington departments of computer science and engineering and biochemistry.

FOURSQUARE

(Chapter 8) Sign up for this social life–management game at www.foursquare .com, or search your smart phone's app store or market for the Foursquare mobile phone app.

FREE RICE

(Chapter 11) Play games to help end hunger at http://freerice.com, a nonprofit website run by the United Nations World Food Programme.

GHOSTS OF A CHANCE

(Chapter 9) Explore the archive of the Smithsonian Museum's experimental game and sign up to play a ninety-minute version at the museum at www .ghostsofachance.com.

GROUNDCREW

(Chapter 12) Organize your own team of agents to tackle any social problem at http://groundcrew.us, and learn more about the company behind the platform and its founder, Joe Edelman, at http://citizenlogistics.com.

HIDE & SEEK FESTIVAL AND SANDPIT

(Chapter 9) Keep track of new mobile, social immersive experiences and games being invented and publicly playtested in the United Kingdom at www .hideandseekfest.co.uk.

INVESTIGATE YOUR MP'S EXPENSES

(Chapter 11) Play with the crowdsourcing tool and read updates about the *Guardian*'s political investigation of UK parliament members at http:// mps-expenses.guardian.co.uk.

JETSET

(Chapter 8) See more screenshots from the airport game developed by Persuasive Games, and download it for your iPhone, at www.persuasivegames .com/games/game.aspx?game=jetset.

LOST JOULES

(Chapter 12) Learn more about this pending smart-meter game project, powered by Adaptive Meter, at http://lostjoules.com.

THE LOST RING

(Chapter 13) To learn more about The Lost Ring, created as a partnership between McDonald's, AKQA, Jane McGonigal, and the International Olympic Committee, explore the player-created wiki at http://olympics.wiki

bruce.com/Home or watch the interactive case study at http://work.akqa.com/thelostring/.

NIKE+

(Chapter 8) View all the Nike+ challenges and sign up to join the running game at www.nikeplus.com; an inexpensive Nike+ sensor and an iPhone or iPod are required to play.

PLUSONEME

(Chapter 8) Send someone you know a +1 inspiring, +1 kindness, +1 humor, or dozens of other positive strengths at http://plusoneme.com, created by Clay Johnson.

QUEST TO LEARN

(Chapter 7) Download sample curricula and assignments at the world's first game-based public school, developed by the Institute of Play, at http://q2l.org.

SPORE CREATURE CREATOR

(Chapter 13) Contribute to the *Spore* galaxy by creating your own *Spore* creature for free at Maxis/Electronic Arts' www.spore.com.

SUPERBETTER

(Chapter 7) To learn the rules for this injury or illness recovery game, or to share them with friends and family, visit www.superbetter.org.

SUPERSTRUCT

(Chapter 14) To view the archive of this future-forecasting game, created by the Institute for the Future, go to www.superstructgame.org. To learn more about the game and download game results, visit the Superstruct blog archive at http://iftf.org/search/node/superstruct.

TOMBSTONE HOLD 'EM

(Chapter 10) To learn the rules of this cemetery poker game, invented by Jane McGonigal for 42 Entertainment, visit www.avantgame.com/tombstone.

TOP SECRET DANCE-OFF

(Chapter 10) You can view the first prototype of this dance adventure game, developed by Jane McGonigal, at http://topsecret.ning.com. Future versions of the game will be announced at www.realityisbroken.org.

WORLD WITHOUT OIL

(Chapter 14) Explore a week-by-week replay of the peak-oil simulation, presented by ITVS and produced by Writer Guy, or download lesson plans for teachers, at http://worldwithoutoil.org.

Practical Advice for Gamers

Reality Is Broken explains the science behind why games are good for us—why they make us happier, more creative, more resilient, and better able to lead others in world-changing efforts.

But some games are better for us than others, and there can be too much of a good thing. Here are a few practical guidelines to help you (or the gamer in your life) get the most positive impact from playing games.

This practical advice—five key guidelines, plus two quick rules—is scientifically backed and can be summed up in a single sentence:

Play games you enjoy **no more than twenty-one hours a week**; **face-to-face** with **friends and family** as often as you can; and in **cooperative** or **creator** modes whenever possible.

1. Don't play more than twenty-one hours a week.

Studies show that games benefit us mentally and emotionally when we play up to three hours a day, or twenty-one hours a week. (In extremely stressful circumstances—such as military deployment—research shows that gamers can benefit from as many as twenty-eight hours a week.[1]) But for virtually everyone else, whenever you play more than twenty-one hours a week, the benefits of gaming start to decline sharply. (This is especially true for children and adolescents.[2])

By the time you're spending forty hours or more a week playing games, the psychological benefits have disappeared entirely and are replaced with negative impacts on your physical health, relationships, and real-life goals. So always strive to keep your gaming in the sweet spot: seven to twenty-one hours a week.

2. Playing with real-life friends and family is better than playing alone all the time, or with strangers.

Gaming strengthens your social bonds and builds trust, two key factors in any positive relationship. And the more positive relationships you have in real life, the happier, healthier, and more successful you are.[3]

You can get mental and emotional benefits from single-player games, or by playing with strangers online—but to really unlock the power of games, it's important to play them with people you really know and like as often as possible.

A handy rule of thumb: try to make half of your gaming social. If you play ten hours a week, try to play face-to-face with real-life friends or family for at least five of those hours.

And if you're not a gamer yourself but you have a family member who plays games all the time, it would do you both good to play together. (Even if you think you don't like games!)

3. Playing face-to-face with friends and family beats playing with them online.

If you're in the same physical space, you'll supercharge both the positive emotional impacts and the social bonding.

Many of the benefits of games are derived from the way they make us feel—and all positive emotions are heightened by face-to-face interaction. Plus, research shows that social ties are strengthened much more when we play games in the same room than when we play games together online. This is especially true for parent-child relationships. In particular, daughters who play video games at home with their parents report feeling much closer to their parents—and demonstrate significantly lower levels of aggression, behavior problems, and depression.[4]

Multiplayer games are great for this. But single-player works too! You can get all the same benefits by taking turns at a single-player game, helping and cheering each other on.

4. Cooperative gameplay, overall, has more benefits than competitive gameplay.

Cooperative gameplay lifts our mood longer, and strengthens our friendships more, than competing against each other. Cooperative gameplay also makes us more likely to help someone in real life, and better collaborators at work—boosting our real-world likeability and chances for success.

Competition has its place, too, of course—we learn to trust others and often motivate ourselves to achieve more when we compete. Not to mention, of course, that all games are fundamentally cooperative: Even if we're trying to beat someone, we're cooperating to play by the same rules and to complete the game without quitting.

But if we spend all our time competing with others, especially if we're competing online against strangers, we may experience feelings of aggression and hostility that offset the positive emotions of gaming.[5] So whenever you're gaming with others, be sure to check to see if there are co-op missions or a co-op mode available. An hour of co-op a week goes a long way. You can find great co-op games for every platform, and a family friendly list too, at Co-Optimus (www.co-optimus.com), the best online resource for co-op gaming.

5. Creative games have special positive impacts.

Many games encourage or even require players to design and create as part of the gameplay process—for example *Spore*, *Little Big Planet*, and *Minecraft*; or the *Halo* level designer and the *Guitar Hero* song creator. These games have been shown to build up players' sense of creative agency—and they make us more likely to create something outside of the game. If you want to really build up your own creative powers, creative games are a great place to start.

Of course, you can always take the next creative step and start making your own games. If you've never made a game, it's easier than you think—and there are some great books to help you get started: *The Art of Game Design* by Jesse Schell, *Challenges for Game Designers* by Brenda Brathwaite, *Game Design Workshop* by Tracy Fullerton, *Level Up! The Guide to Great Videogame Design* by Scott Rogers, and *Game Programming for Teens* by Maneesh Sethi. Additional online resources can be found at www.gamasutra.com.

TWO OTHER IMPORTANT RULES:

You can get all of the benefits of a good game without realistic violence—you (or your kids) don't have to play games with guns or gore.

If you feel strongly about violence, look to games in other genres—there's no shortage of amazing sports, music, racing, puzzle, role-playing, casual, strategy, and adventure games. You might also have a rule in your house: No games where the gameplay involves hurting human characters. (I personally only shoot zombies or aliens in games—I just don't like the way realistic violence against humans in a game makes me feel.)

Any game that makes you feel bad is no longer a good game for you to play.

This should be obvious, but sometimes we get so caught up in our games that we forget they're supposed to be fun. If you find yourself feeling really upset when you lose a game, or if you're fighting with friends or strangers when you play, you're too invested. Switch to a different game for a while, a game that has "lower stakes" for you personally.

Or, especially if you play with strangers online, you might find yourself surrounded by other players who say things that make you uncomfortable or who just generally act like jerks. (This happens more often in an online game environment where players can be "anonymous" to each other.) Obnoxious behavior by your fellow players will actually make it harder for you to get the positive benefits of games—so don't waste your time playing with a community that gets you down.

Meanwhile, if you start to wonder if you're spending too much time on a particular game—maybe you're starting to feel just a tiny bit addicted—keep track of your gaming hours for one week. Make sure they add up to less than twenty-one hours! And you may want to limit yourself to even fewer for a little while if you're feeling too much "gamer regret."

Keep these five guidelines and two rules in mind, and you're sure to minimize the potential downsides and maximize the positive impacts of playing games.

Notes

1. Suits, Bernard. *The Grasshopper: Games, Life and Utopia* (Ontario, Canada: Broadview Press, 2005), 159.

INTRODUCTION

1. Castronova, Edward. *Exodus to the Virtual World* (New York: Palgrave Macmillan, 2007), xiv–xvii. This condensed passage from the preface appears with the permission of the author.
2. The largest wiki resource for *World of Warcraft* is WoWWiki (wowwiki.com) and its spin-off wiki, the WoWpedia (wowpedia.org). Wiki page statistics for these wikis and all of the world's biggest wikis, including Wikipedia, are available at http://meta.wikimedia.org/wiki/List_of_largest_wikis.
3. "China Bars Use of Virtual Money for Trading in Real Goods." Press release from the Ministry of Commerce, People's Republic of China, June 29, 2009. http://english.mofcom.gov.cn/aarticle/news release/commonnews/200906/20090606364208.html.
4. "Newzoo Games Market Report: Consumer Spending in US, UK, GER, FR, NL, & BE." Newzoo, Amsterdam, May 2010. http://corporate.newzoo.com/press/GamesMarketReport_FREE_030510 .pdf; "Games Segmentation 2008 Market Research Report." The NPD Group, May 2010. http://www.npd.com/press/releases/press_100527b.html.
5. These regional statistics are drawn from a variety of industry reports and market research studies conducted over the past three years, as follows: "An Analysis of Europe's 100 Million Active Gamers." Strategy Analytics, September 2008. http://www.strategyanalytics.com/reports/ix7hx8in7j/single.htm; "IA9 Interactive Australia 2009." National research prepared by Bond University for Interactive Games & Entertainment Association, August 2009. www.igea.net/wp.../2009/08/IA9-Interactive-Australia-2009-Full-Report.pdf; "Online Games Market in Korea." Pearl Research, July 2009. http://www.researchandmarkets.com/reportinfo.asp?report_id=1208384; "Games Market Summary: Russia." Prepared by Piers Harding-Rolls for Games Intelligence/Screen Digest, June 2010. http://www.screendigest.com/intelligence/games/russia/games_intel_russia_100/view.html?start_ser=gi&start_toc=1; "Emerging Markets for Video Games." Chris Stanton-Jones for Games Intelligence/Screen Digest, March 2010. http://www.screendigest.com/reports/10_03_18_emerging_markets_video_games/10_03_18_emerging_markets_video_games/view.html; "Online Games Market in Vietnam." Pearl Research, November 2008. http://www.mindbranch.com/Online-Games-Vietnam-R740-14/; "Study: Vietnam, India Gaming Population To Hit 25 Million By 2014." Pearl Research, March 2010. http://www.gamasutra.com/view/news/27525/Study_Vietnam_India_Gaming_Population_

To_Hit_25Million_By_2014.php; "Gaming Business in the Middle East." Game Power 7 Research Group, February 2010. http://images.bbgsite.com/news/download/gaming_business_in_the_ middle_east_KOGAL_2009.pdf; Menon, Naveen. "Insights on Mobile Gaming in India." Vital Analytics. March 2009. http://www.telecomindiaonline.com/telecom-india-daily-telecom-insights-on-mobile-gaming-in-urban-india.html. "The Global Entertainment & Media Outlook: 2010-2014." PricewaterhouseCoopers, June 2010. http://www.pwc.com/gx/en/global-entertainment-media-out-look. The gamer market statistics are constantly changing; the global trends are upward in every country, so higher numbers are likely to be accurate for each region with each passing year since the studies were published.

6. "Major Findings of the 2008 Annual Review & Five-Year Forecast Report on China's Video Game Industry." Niko Partners Research, San Jose, May 2, 2008. www.nikopartners.com/pdf/npr_050208 .pdf.

7. "Games Segmentation 2008 Market Research Report." The NPD Group.

8. Dromgoole, Sean. "A View from the Marketplace: Games Now and Going Forward." GameVision Europe Ltd., March 2009. http://www.scribd.com/doc/13714815/Sean-Dromgoole-CEO-Some-Research-Gamevision.

9. In 2009, the annual spending on games in the United States was $25.3 billion; in the United Kingdom, it was £3.8 billion; in Germany, 3.7 billion euros; and in France, 3.6 billion euros. "Newzoo Games Market Report."

10. Rawlinson, George, trans., with Henry Rawlinson and J. G. Wilkinson. *The History of Herodotus: A New English Version* (New York: D. Appleton, 1861), 182. http://www.archive.org/stream/ historyofherodot01herouoft#page/n5/mode/2up.

11. The first such taxes have already been proposed; for example, in Texas in 2006 and by lawmakers in New Mexico in 2008. Reported in "Texas Politician Proposes 100 Percent Game Tax." GameSpot, January 25, 2006. http://www.gamespot.com/news/6143114.html; and "New Mexico's Videogame Nanny Tax." CNET News, February 11, 2008. http://news.cnet.com/New-Mexicos-video-game-nanny-tax/2010-1043_3-6229759.html.

12. "Essential Facts About the Game Industry: 2010 Sales, Demographic and Usage Data." Entertainment Software Association, June 16, 2010. http://www.theesa.com/facts/pdfs/ESA_Essential_Facts_ 2010.PDF.

13. Reinecke, Leonard. "Games at Work: The Recreational Use of Computer Games During Work Hours." *Cyberpsychology, Behavior, and Social Networking* [formerly *CyberPsychology & Behavior*], August 2009, 12(4): 461–65. DOI: 10.1089/cpb.2009.0010.

14. Fahey, Rob. "It's Inevitable: Soon We Will All Be Gamers." *The Times* (UK), July 7, 2008. http:// www.timesonline.co.uk/tol/comment/columnists/guest_contributors/article4281768.ece.

PART I

1. Csíkszentmihályi, Mihály. *Beyond Boredom and Anxiety: The Experience of Play in Work and Games* (San Francisco: Jossey-Bass, 1975), 206.

CHAPTER 1

1. Suits, *The Grasshopper*, 38. Katie Salen and Eric Zimmerman were among the first game researchers to outline these three characteristics as essential to a game, drawing on the work of Bernard Suits. I am indebted to them, as are many other game designers and researchers, for drawing attention to Suits' definition. Salen, Katie, and Eric Zimmerman. *Rules of Play: Game Design Fundamentals* (Cambridge: MIT Press, 2004).

2. *Tetris* has been named "The Greatest Game of All Time" by *Electronic Gaming Monthly*, Issue 100, among others. "The Best Videogames in the History of Humanity," an extraordinary compilation of greatest-games lists compiled by J. J. McCullough, can be found at http://www.filibustercartoons. com/games.htm.

3. A master's thesis written proving that *Tetris* is literally unwinnable: Brzustowski, John. "Can You Win at *Tetris?*" University of British Columbia, Master of Science, Mathematics, 1992. http://www.iam .ubc.ca/theses/Brzustowski/brzustowski.html.

4. Csíkszentmihályi, *Beyond Boredom and Anxiety*, 36.

5. Carse, James P. *Finite and Infinite Games: A Vision of Life as Play and Possibility* (New York: Free Press, 1986), 3.

6. Sutton-Smith, Brian. *Ambiguity of Play* (Cambridge: Harvard University Press, 2001), 198.

7. This kind of real-time data collection helps researchers collect much better data than traditional surveys or questionnaires. It's much easier to report your activity and mood in the moment than to try to remember what you were doing and feeling hours or days afterward.

8. Csíkszentmihályi, Mihály. *Flow: The Psychology of Optimal Experience* (New York: Harper Perennial, 1991). See also: Csíkszentmihályi, M. "Optimal Experience in Work and Leisure." *Journal of Personality and Social Psychology*, 1989, 56(5): 815–22; Csíkszentmihályi, M., and R. Kubey. "Television and the Rest of Life: A Systematic Comparison of Subjective Experience." *Public Opinion Quarterly*, 1981, 45: 317–28; and Kubey, R., R. Larson, and M. Csíkszentmihályi. "Experience Sampling Method Applications to Communication Research Questions." *Journal of Communication*, 1996, 46(2): 99–120.

9. Kash, Thomas, et al. "Dopamine Enhances Fast Excitatory Synaptic Transmission in the Extended Amygdala by a CRF-R1-Dependent Process." *Journal of Neuroscience*, December 17, 2008, 28(51): 13856–65. DOI: 10.1523/JNEUROSCI.4715-08.2008.

10. Gregory, Erik M. "Understanding Video Gaming's Engagement: Flow and Its Application to Interactive Media." *Media Psychology Review*, Issue 1, 2008. http://www.mprcenter.org/mpr/index .php?option=com_content&view=article&id=207&Itemid=163.

11. Yee, Nick. "MMORPG Hours vs. TV Hours." The Daedalus Project: The Psychology of MMORPGS, March 9, 2009, vol. 7–1. http://www.nickyee.com/daedalus/archives/000891.php.

12. Ben-Shahar, Tal. *Happier: Learn the Secrets to Daily Joy and Lasting Fulfillment* (New York: McGraw-Hill, 2007), 77.

13. Nicole Lazzaro deserves the credit for introducing this term to the game industry via her presentations at the annual Game Developers Conference.

14. Hoeft, Fumiko, et al. "Gender Differences in the Mesocorticolimbic System During Computer Game-Play." *Journal of Psychiatric Research*, March 2008, 42(4): 253–8. http://spnl.stanford.edu/publications/pdfs/Hoeft_2008JPsychiatrRes.pdf.

CHAPTER 2

1. Csíkszentmihályi, *Beyond Boredom and Anxiety*, xiii.

2. Ibid., 37.

3. Ibid., 1, 197.

4. "Study: Women Over 40 Biggest Gamers." CNN, February 11, 2004. http://edition.cnn.com/2004/TECH/fun.games/02/11/video.games.women.reut/index.html.

5. Sutton-Smith, Brian, and Elliot Avedon, eds. *The Study of Games* (New York: Wiley, 1971).

6. Thompson, Clive. "*Halo 3*: How Microsoft Labs Invented a New Science of Play." *Wired*, August 21, 2007. http://www.wired.com/gaming/virtualworlds/magazine/15-09/ff_halo.

7. Sudnow, David. *Pilgrim in the Microworld* (New York: Warner Books, 1983), 41. Full text available online at http://www.sudnow.com/PMW.pdf.

8. Ibid., 35.

9. Ibid., 9.

10. Ibid., 20.

11. Corey Lee M. Keyes explains how flow fits into the bigger picture of positive psychology in "What Is Positive Psychology?" CNN, January 24, 2001. http://archives.cnn.com/2001/fyi/teachers.tools/01/24/c.keyes/.

12. Hoeft, Fumiko, et al., "Gender Differences."

13. Thompson, Clive. "Battle with 'Gamer Regret' Never Ceases." *Wired*, September 10, 2007. http://www.wired.com/gaming/virtualworlds/commentary/games/2007/09/gamesfrontiers_0910?current Page=all.

14. Jenkins, David. "Chinese Online Publishers Sign 'Beijing Accord.'" Gamasutra News, August 24, 2005. http://www.gamasutra.com/php-bin/news_index.php?story=6312.

15. Lyubomirsky, S., K. M. Sheldon, and D. Schkade. "Pursuing Happiness: The Architecture of Sustainable Change." *Review of General Psychology*, 2005, 9: 111–31; and Sheldon, K. M., and S. Lyubomirsky. "Is It Possible to Become Happier? (And if So, How?)" *Social and Personality Psychology Compass*, 2007, 1: 1–17.

16. Originally described by Brickman and Campbell in "Hedonic Relativism and Planning the Good Society." In M. H. Apley, ed., *Adaptation Level Theory: A Symposium* (New York: Academic Press, 1971), 287–302. Most recently assessed in Bottan, Nicolas Luis, Pérez Truglia, and Ricardo Nicolás. "Deconstructing the Hedonic Treadmill: Is Happiness Autoregressive?" Social Science Research Network, January 2010. http://ssrn.com/abstract=1262569.

17. The term "autotelic" was originally coined by Csíkszentmihályi in *Beyond Boredom and Anxiety*, 10.

18. Lyubomirsky, Sonja. *The How of Happiness: A Scientific Approach to Getting the Life You Want* (New York: Penguin Press, 2008), 64.

19. Nelson, Debra L., and Bret L. Simmons. "Eustress: An Elusive Construct, and Engaging Pursuit." *Research in Occupational Stress and Well-being*, 2003, 3: 265–322.

20. I owe credit to Chris Bateman for enlightening me about the neurochemical basis for fiero, in "Top Ten Videogame Emotions." Only a Game, April 9, 2008. http://onlyagame.typepad.com/only_a_game/2008/04/top-ten-videoga.html.

21. Berns, G. S. "Something Funny Happened to Reward." *Trends in Cognitive Sciences*, 2004, 8(5): 193–94. DOI: 10.1016/j.tics.2004.03.007.

22. Keltner, Dacher. *Born to Be Good: The Science of a Meaningful Life* (New York: Norton, 2009), 219–20.

23. Ibid., 250–69.

24. Gilbert, Elizabeth. *Eat, Pray, Love* (New York: Viking, 2006), 260.

25. Many of the studies whose results I've based this set of intrinsic rewards on are reviewed in the following books: Martin Seligman's *Authentic Happiness* and *Learned Optimism*; Seligman and Christopher Peterson's *Character Strengths and Virtues*; Sonja Lyubomirsky's *The How of Happiness*; Tal Ben-Shahar's *Happier*; Jean M. Twenge's *Generation Me*; Mihály Csíkszentmihályi's *Beyond Boredom and Anxiety*; and Eric Weiner's *The Geography of Bliss*.

26. Lyubomirsky, *The How of Happiness*, 16.

CHAPTER 3

1. That's a back-of-the-envelope calculation—derived from adding up the average number of subscribers each year since 2004, ranging from 2 million to 11.5 million, times the average hours per player according to Blizzard statistics and Stanford University research, added up through early 2010. This isn't a precise way to measure gameplay, but even with a margin of error of as much as 50 percent, we're still talking about gameplay on the magnitude of millions of years.

2. In 2008, scientists discovered fossils suggesting man's earliest upright ancestors date to 6 million years ago. Richmond, Brian G., and William L. Jungers. "*Orrorin tugenensis* Femoral Morphology and the Evolution of Hominin Bipedalism." *Science*, March 21, 2008, 319(5870): 1662. DOI: 10.1126/science.1154197.

3. Nielsen reports seventeen hours per week per *World of Warcraft* player, while a Stanford researcher reports twenty-two hours; other studies confirm consistent numbers in this range. "Online Games Battle for Top Spot." BBC News, December 26, 2007. http://news.bbc.co.uk/2/hi/technology/7156078.stm; and "WoW Basic Demographics." The Daedalus Project, 2009, vol. 7–1. http://www.nickyee.com/daedalus/archives/001365.php.

4. "Full Transcript of Blizzard 2010 Plans." Activision Blizzard Fourth-Quarter Earnings, conference

call, posted online by INC Gamers, February 11, 2010. http://www.incgamers.com/News/20949/ full-transcript-blizzard-2010-plans; and "World of Warcraft Expansion Shatters Sales Records." PC Magazine, November 20, 2008. http://www.pcmag.com/article2/0,2817,2335141,00.asp.

5. The term "blissful productivity" was first applied to WoW by a team of computer scientists at the Indiana University School of Informatics; they were studying the unusually high gaming stamina of WoW players, particularly in the face of what seemed like repetitive and often tedious tasks. Bardzell, S., J. Bardzell, T. Pace, and K. Reed. "Blissfully Productive: Grouping and Cooperation in World of Warcraft Instance Runs." In Proceedings of the 2008 ACM Conference on Computer Supported Co-operative Work, San Diego, November 8–12, 2008 (New York: ACM, 2008), 357–60. DOI: http:// doi.acm.org/10.1145/1460563.1460621.

6. Castronova, Exodus to the Virtual World, 124.

7. Cavalli, Earnest. "Age of Conan's Maximum Level Only 250 Hours Away." Wired, May 14, 2008. http://www.wired.com/gamelife/2008/05/age-of-conans-m/.

8. "Why Questing Is the Fastest and Most Enjoyable Way to Level." WoW Horde Leveling, January 23, 2007. http://wowhordeleveling.blogspot.com/2007/01/why-questing-is-fastest-and-most.html.

9. Cavalli, "Age of Conan's Maximum Level."

10. Lyubomirsky, The How of Happiness, 67.

11. "World of Warcraft: Wrath of Lich King Review." GameSpot, November 13, 2008. http://www .gamespot.com/pc/rpg/worldofwarcraftwrathofthelichking/review.html.

12. Seligman, Martin. Authentic Happiness (New York: Free Press, 2004), 40.

13. "Raiding for Newbies." WoWWiki. http://www.wowwiki.com/Raiding_for_newbies.

14. "World of Warcraft: Phasing Explained." MMORPG. September 28, 2008. http://www.mmorpg .com/discussion2.cfm/post/2329941#2329941.

15. Calvert, Justin. "Wrath of Lich King Review." Gamespot. November 13, 2008. http://www.gamespot .com/pc/rpg/worldofwarcraftwrathofthelichking/review.html.

16. De Botton, Alain. The Pleasures and Sorrows of Work (New York: Pantheon, 2009), 80.

17. Crawford, Matthew. "The Case for Working with Your Hands." New York Times Magazine, May 21, 2009. http://www.nytimes.com/2009/05/24/magazine/24labor-t.html.

18. Reinecke, Leonard, "Games at Work."

19. De Botton, The Pleasures and Sorrows of Work, 260.

CHAPTER 4

1. Personal interview with Nicole Lazzaro, April 25, 2009.

2. Ravaja, Niklas, Timo Saari, Jari Laarni, Kari Kallinen, and Mikko Salminen. "The Psychophysiology of Video Gaming: Phasic Emotional Responses to Game Events." Changing Views: World in Play. Digital Games Research Association, June 2005. http://www.digra.org/dl/db/06278.36196.pdf.

3. The study was named #1 Finding in Games Research at the annual Game Developers Conference in 2006. References for all of the top studies can be found at http://www.avantgame.com/top10.htm.

4. "Can I Have My Life Back?" Player review of Super Monkey Ball, Amazon, July 1, 2002. http://www .amazon.co.uk/review/R27IJK4R3ITHIR/ref=cm_cr_rdp_perm/.

5. Thompson, Clive. "The Joy of Sucking." Wired, July 17, 2006. http://www.wired.com/gaming/gam ingreviews/commentary/games/2006/07/71386?currentPage=all.

6. Koster, Raph. A Theory of Fun for Game Design (Phoenix: Paraglyph Press, 2004), 8–9.

7. Ibid., 40.

8. Ibid., 118.

9. Seligman, Martin. Learned Optimism: How to Change Your Mind and Your Life (New York: Free Press, 1998), 164–66.

10. Ibid., 69.

11. Nesse, R. M. "Is Depression an Adaptation?" Archives of General Psychiatry, 2000, 57: 14–20. Full text also available at http://www-personal.umich.edu/~nesse/Articles/IsDepAdapt-ArchGenPsy-chiat-2000.pdf; and Kellera, M. C., and R. M. Nesse. "The Evolutionary Significance of Low Mood

Symptoms." *Journal of Personality and Social Psychology*, 2005, 91(2): 316–30. Full text also available at http://www-personal.umich.edu/~nesse/Articles/Keller-Nesse-EvolDepSx-JPSP-2006.pdf.

12. Lyubomirsky, *The How of Happiness*, 213.

13. "*Rock Band* Franchise Officially Surpasses $1 Billion in North American Retail Sales, According to the NPD Group." Company press release, New York, March 26, 2009. http://www.rockband.com/news/one_billion_dollars.

14. Davies, Chris. "Pro Drum Kit Mod into Full-Size *Rock Band* Controller." Slash Gear, January 11 2009. http://www.slashgear.com/pro-drum-kit-mod-into-full-size-rock-band-controller-119585/.

15. "*Guitar Hero II*: Playing vs. Performing a Tune." Ludologist. http://www.jesperjuul.net/ludologist/?p=312; and "In *Rock Band*, Actually Play Drums and Sing." Ludologist. http://www.jesperjuul.net/ludologist/?p=412.

16. Lang, Derrik J. "*Rock Band 2* Will Include New Instruments, Online Modes, Songs." Associated Press, June 30, 2008. Accessible at http://www.usatoday.com/tech/gaming/2008-06-30-rock-band-2_N.htm.

17. Brightman, James. "*Guitar Hero, Rock Band* Players Showing Increased Interest in Real Instruments." GameDaily, November 25, 2008. http://www.gamedaily.com/games/rock-band-2/playstation-3/game-news/guitar-hero-rock-band-players-showing-increased-interest-in-real-instruments/.

18. Seligman, *Learned Optimism*, 174.

CHAPTER 5

1. The gameplay is so similar, in fact, Lexulous (formerly known as Scrabulous) barely survived a copyright infringement lawsuit by the creators of the original board game. Timmons, Heather. "Scrabble Tries to Fight a Popular Impostor at Its Own Game." *New York Times*, April 7, 2008. http://www.nytimes.com/2008/04/07/technology/07scrabulous.html.

2. McDonald, Thomas. "Absolutely Scrabulous!" *Maximum PC*, September 24, 2008. http://www.maximumpc.com/article/%5Bprimary-term%5D/absolutely_scrabulous.

3. "What Happened to Scrabulous?" New Home of Suddenly Susan, September 26, 2008. http://desperatelyseekingsuddenlysusan.wordpress.com/2008/09/26/what-happened-to-scrabulous/.

4. For example: "Facebook Friends?," screenshot taken March 27, 2009. http://www.flickr.com/photos/bennynerd/3389278659/; "Baby's First Scrabulous Game," screenshot taken February 28, 2008. http://www.flickr.com/photos/chickitamarie/2299675218/; and "Online Scrabble with Mom," screenshot taken April 15, 2009. http://www.flickr.com/photos/jaboney/3444811350/.

5. "Domination," screenshot taken May 14, 2008. http://www.flickr.com/photos/yummiec00kies/2494160470/.

6. "The Big Pic+ure of my #U+$@(< Kicking," screenshot taken May 27, 2008. http://www.flickr.com/photos/hemantvt83/2529818600/.

7. "Stepdaughter Spurns Scheduled Scrabulous." Postcards from Yo Mama, February 19, 2009. http://www.postcardsfromyomomma.com/2009/02/19/stepdaughter-spurns-scheduled-scrabulous/.

8. "Loving Scrabulous," screenshot taken August 12, 2007. http://www.flickr.com/photos/etches-johnson/1095923577/.

9. "Funny Lexulous Game," screenshot taken July 11, 2009. http://www.flickr.com/photos/avantgame/3710408343/in/photostream/.

10. "Bring It, Ben!," screenshot taken April 10, 2008. http://www.flickr.com/photos/kendalchen/2404592798/.

11. Brophy-Warren, Jamin. "Networking Your Way to a Triple-Word Score." *Wall Street Journal*, October 13, 2007. http://online.wsj.com/public/article_print/SB119222790761657777.html.

12. Weiner, Eric. *The Geography of Bliss* (New York: Twelve, 2008), 325.

13. Ibid., 114.

14. McElroy, Griffin. "FarmVille Community Surpasses 80 Million Players." Joystiq, February 20, 2010. http://www.joystiq.com/2010/02/20/farmville-community-surpasses-80-million-players/; and FarmVille Application Info, AppData, accessed March 2010. Current statistics available at http://www.appdata.com/facebook/apps/index/id/102452128776.

15. "The Most Intense Game of Scrabulous Ever," screenshot taken June 3, 2008. http://www.flickr .com/photos/mariss007/2547926935/.

16. "Online Scrabble with Mom."

17. "My Amazing Lexulous Score—87 points!," screenshot taken June 12, 2009. http://www.flickr.com/ photos/sour_patch/3621419260/.

18. "Pwn." Wikipedia entry, accessed May 1, 2010. http://en.wikipedia.org/wiki/Pwn.

19. Keltner, *Born to Be Good*, 163.

20. "*WarioWare: Smooth Moves* Review." GameSpot, January 12, 2007. http://www.gamespot.com/wii/ puzzle/wariowaresmoothmoves/review.html.

21. Bateman, "Top Ten Videogame Emotions."

22. Ibid.

23. Ekman, Paul. *Emotions Revealed: Recognizing Faces and Feelings to Improve Communication and Emotional Life* (New York: Times Books, 2003), 197.

24. Seligman, *Learned Optimism*, 282.

25. Ibid., 282, 284.

26. Jenkins, Henry. "Reality Bytes: Eight Myths about Videogames Debunked." The Videogame Revolution, PBS, 2005. http://www.pbs.org/kcts/videogamerevolution/impact/myths.html.

27. "*Braid* Review." IGN, August 4, 2008. http://xboxlive.ign.com/articles/896/896371p1.html.

28. "*Braid* Is Now Live." Official *Braid* blog, August 5, 2008. http://braid-game.com/news/?p=255.

29. "*Braid* Thrives on Live." *Edge*, August 13, 2008. http://www.edge-online.com/news/braid-thrives-live.

30. Ducheneaut, Nicolas, Nicholas Yee, Eric Nickell, and Robert J. Moore. "Alone Together? Exploring the Dynamics of Massively Multiplayer Online Games." In Conference Proceedings on Human Factors in Computing Systems, CHI 2006, Montreal, Canada, April 22–27, 2006, 407–16. http:// www.nickyee.com/pubs/Ducheneaut,%20Yee,%20Nickell,%20Moore%20-%20Alone%20To-gether%20(2006).pdf.

31. Short, J., E. Williams, and B. Christie. *The Social Psychology of Telecommunications* (London: Wiley, 1976).

32. Morrill, Calvin, David A. Snow, and Cindy H. White. *Together Alone: Personal Relationships in Public Spaces* (Berkeley: University of California Press, 2005). The term "alone together" in the gaming context is inspired by this social theory text, which describes "social ties that paradoxically blend aspects of durability and brevity, of emotional closeness and distance, of being together and alone."

33. "That's Right! I Solo in Your MMOs!" Mystic Worlds, June 9, 2009. http://notadiary.typepad.com/ mysticworlds/2009/06/thats-right-i-solo-in-your-mmos.html.

34. Ibid.

35. Myers, David G. "The Secrets of Happiness." *Psychology Today*, October 2009. http://www.psy chologytoday.com/articles/199207/the-secrets-happiness?page=2.

36. Ito, Mizuko, et al. *Hanging Out, Messing Around, and Geeking Out: Kids Living and Learning with New Media* (Cambridge: MIT Press, 2009), 195.

37. Cookman, Daniel. "Pick Your Game Community: Virtual, or Real?" Lost Garden, February 5, 2006. http://lostgarden.com/2006/02/pick-your-game-community-virtual-or.html.

38. Ibid.

39. "12 Ways Video Games Actually Benefit 'Real Life.'" Pwn or Die, May 12, 2009. http://www.pwnor die.com/blog/posts/15739.

CHAPTER 6

1. "13 Billion Kills: Join the Mission." *Halo 3* forum, Bungie.net, February 20, 2009. http://www .bungie.net/News/content.aspx?type=topnews&link=TenBillionKills.

2. Ibid.

3. "Players Attempt to Hit 7 Billion Kills While *Halo 3* Killcount Exceeds Global Population." Joystiq, June 27, 2008. http://xbox.joystiq.com/2008/06/27/players-attempt-to-hit-7-billion-kills-while-halo-3-killcount-ex/.

4. Leith, Sam. "*Halo* 3. Blown Away." *Telegraph*, September 22, 2007. http://www.telegraph.co.uk/culture/3668103/Halo-3-blown-away.html.

5. "Campaign Kill Count: 10,000,000,000." *Halo* 3 forum, Bungie.net, April 13, 2009. http://www.bungie.net/Forums/posts.aspx?postID=32064021&postRepeater1-p=3.

6. Ibid.

7. Ibid.

8. Seligman, *Learned Optimism*, 287.

9. "Bungie: 10 Billion Covenant Killed in *Halo* 3 . . . and Growing." Joystiq, April 13, 2009. http://xbox.joystiq.com/2009/04/13/bungie-10-billion-covenant-killed-in-halo-3-and-growing/.

10. "*Halo* 3 Review." NZGamer, September 24, 2007. http://nzgamer.com/x360/reviews/538/halo-3.html.

11. Paul Pearsall. *Awe: The Delights and Dangers of Our Eleventh Emotion* (Deerfield Beach, Florida: HCI, 2007), 193.

12. Keltner, *Born to Be Good*, 268.

13. Polack, Trent. "Epic Scale." Gamasutra, July 16, 2009. http://www.gamasutra.com/blogs/Trent Polack/20090716/2412/Epic_Scale.php.

14. Kuhrcke, Tim, Christoph Klimmt, and Peter Vorderer. "Why Is Virtual Fighting Fun? Motivational Predictors of Exposure to Violent Video Games." Paper presented at the annual meeting of the International Communication Association, Dresden, Germany, May 25, 2009. http://www.allacademic.com/meta/p91358_index.html.

15. "Return of the New Hotness." Bungie.net, August 27, 2009. http://www.bungie.net/news/content.aspx?type=topnews&link=NewHotness.

16. Kelly, Kevin. "Scan This Book!" *New York Times*, March 14, 2006. http://www.nytimes.com/2006/05/14/magazine/14publishing.html.

17. "Watch New *Halo* 3 Ad: 'Two Soldiers Reminisce.'" Joystiq, September 22, 2007. http://www.joystiq.com/2007/09/22/watch-new-halo-3-ad-two-soldiers-reminisce/.

18. "*Halo* 3 Ad Brings Battle to Reality." *Escapist*, September 12, 2007. http://www.escapistmagazine.com/forums/read/7.48542.

19. Crecente, Brian. "*Halo* Diorama May Tour Country." Kotaku, September 13, 2007. http://kotaku.com/gaming/gallery/halo-diorama-may-tour-country-299470.php.

20. "Watch New *Halo* 3 Ad," Joystiq.

21. "Hindsight: *Halo* 3." Ascendant Justice, March 1, 2008. http://blog.ascendantjustice.com/halo-3/hindsight-halo-3/.

22. Leith, "*Halo* 3: Blown Away."

23. Curry, Andrew. "Gobekli Tepe: The World's First Temple?" *Smithsonian*, November 2008. http://www.smithsonianmag.com/history-archaeology/gobekli-tepe.html#ixzz0T0oKlRQ6.

24. McIntosh, Lindsay. "'Neolithic Cathedral Built to Amaze' Unearthed in Orkney Dig." *The Times* (UK), August 14, 2009. http://www.timesonline.co.uk/tol/news/uk/scotland/article6795316.ece.

25. Curry, "Gobekli Tepe."

26. Ibid.

27. "Just the Right Sense of 'Ancient.'" Xbox.com, February 19, 2002. www.xbox.com/en-US/games/splash/h/halo/themakers3.htm. Referenced in Wikipedia entry, "Halo Original Soundtrack," accessed May 1, 2010. http://en.wikipedia.org/wiki/Halo_Original_Soundtrack. Originally quoted in http://www.xbox.com/en-US/games/h/halo/themakers3.htm.

28. "NCAA *Football 10* Review." Team Xbox, IGN, July 10, 2009. http://r.views.teamxbox.com/xbox-360/1736/NCAA-Football-10/p1/.

29. Robertson, Margaret. "One More Go: Why *Halo* Makes Me Want to Lay Down and Die." Offworld, September 25, 2009. http://www.offworld.com/2009/09/one-more-go-why-halo-makes-me.html.

30. "*Halo* 3 Wiki: About." HaloWiki, version 22:53, February 13, 2009. http://halowiki.net/p/Main_Page.

31. "NCAA *Football 10* Season Showdown." Inside EA Sports, May 14, 2009. http://insideblog.easports.com/archive/2009/05/14/ncaa-football-10-season-showdown.aspx.

32. Huizinga, Johan. *Homo Ludens* (Boston: Beacon Press, 1971), 446.

33. Gentile, Douglas A., Craig A. Anderson, Shintaro Yukawa, et al. "The Effects of Prosocial Video

Games on Prosocial Behaviors: International Evidence From Correlational, Longitudinal, and Experimental Studies." *Personality and Social Psychology Bulletin*, 2009, 35: 752–63.

34. "Some Video Games Can Make Children Kinder and More Likely to Help." Science Daily, June 18, 2009. http://www.sciencedaily.com/releases/2009/06/090617171819.htm.

35. Maslow, Abraham. *Motivation and Personality* (New York: Harper Collins, 1987), 113.

CHAPTER 7

1. Personal website of Kevan Davis, accessed May 1, 2010. http://kevan.org/cv.

2. "Chore Wars Player Testimonials." Chore Wars. http://www.chorewars.com/testimonials.php.

3. McGonigal, Jane. *This Might Be Game: Ubiquitous Play and Performance at the Turn of the Twenty-First Century*. Dissertation for the PhD in Performace Studies at the University of California, Berkeley. 2006. http://avantgame.com/McGonigal_THIS_MIGHT_BE_A_GAME_sm.pdf

4. For example, the alternate reality game Why So Serious? by 42 Entertainment for *The Dark Knight* achieved an audience of more than 10 million people, according to the viral *Dark Knight* "Why So Serious?" case study. (viewable at http://www.youtube.com/watch?v=cD-HRI-N3Lg). The huge success of this ARG can be attributed not only to excellent game design, but also to the global popularity of the *Batman* movie franchise with which the game was linked.

5. Ito, Mizuko, Heather A. Horst, Matteo Bittanti, Danah Boyd, Becky Herr-Stephenson, Patricia G. Lange, C. J. Pascoe, and Laura Robinson, et al. "Living and Learning with New Media: Summary of Findings from the Digital Youth Project." White paper, The John D. and Catherine T. MacArthur Foundation Reports on Digital Media and Learning, November 2008. http://digitalyouth.ischool. berkeley.edu/report.

6. Prensky, Marc. "Engage Me or Enrage Me: What Today's Learners Demand." *Educause Review*, September/October 2005, 40(5): 60. http://net.educause.edu/ir/library/pdf/erm0553.pdf.

7. King, Nigel S. "Post-Concussion Syndrome: Clarity Amid the Controversy?" *British Journal of Psychiatry*, 2003, 183: 276–78.

8. Thompson, Bronwyn. "Goals and Goal Setting in Pain Management." HealthSkills, December 1, 2008. http://healthskills.wordpress.com/2008/12/01/goals-and-goal-setting-in-pain-management/.

9. McGonigal, Jane. "SuperBetter: Or How to Turn Recovery into a Multiplayer Experience." Avant Game, September 25, 2009. http://blog.avantgame.com/2009/09/super-better-or-how-to-turn-recovery.html.

10. DeKoven, Bernie. "Creating the Play Community." In *The New Games Book* (Garden City, NY: Doubleday, 1976), 41–42.

CHAPTER 8

1. "Key Findings from a Survey of Air Travelers." Peter D. Hart Research Associates/The Winston Group, May 30, 2008. http://www.poweroftravel.org/statistics/pdf/ki_dom_atp_summary.pdf.

2. Seligman, *Learned Optimism*, 17–30.

3. "Jetset: A Serious Game for iPhones." VuBlog, February 16, 2009. http://blog.vudat.msu.edu/?p=232.

4. Day in the Cloud Challenge. Official website, accessed June 24, 2009. http://www.dayinthecloud.com.

5. The correct answer is *The Graduate*, for the phrase "Mrs. Robinson, are you trying to seduce me?"

6. Creative submissions work differently from the puzzles; they don't automatically unlock points while you're on the plane, because they have to be judged for creative merit by game masters on the ground. It's a bit of delayed gratification in its current design, but you could easily imagine an updated version that allows travelers waiting at boarding gates or buying tickets online, for example, to browse and rate submissions so that creative submissions would be rated before the players land.

7. "Day in the Cloud—Virgin Flight 921." Onigame, June 24, 2009. http://onigame.livejournal.com/41979.html.

8. Ibid.

9. "Nike Plus Is a Statwhore Online Game That You Play by Running." Ilxor.com, July 13, 2009. http://www.ilxor.com/ILX/ThreadSelectedControllerServlet?boardid=67&threadid=73699.

10. Fox, Jesse, and Jeremy N. Bailenson. "Virtual Self-Modeling: The Effects of Vicarious Reinforcement and Identification on Exercise Behaviors." *Media Psychology*, 2009, 12: 1–25. DOI: 10.1080/15213260802669474.

11. Donath, Judith. "Artificial Pets: Simple Behaviors Elicit Complex Attachments." In Marc Bekoff, ed., *The Encyclopedia of Animal Behavior* (Santa Barbara, CA: Greenwood Press, 2004).

12. "How Often Do Foursquare Users Actually Check In?" *Business Insider*, May 7, 2010. http://www.businessinsider.com/how-often-do-foursquare-user-actually-check-in-2010-5.

13. "Left the House." Product page on Split Reason. http://www.splitreason.com/product/622.

CHAPTER 9

1. Evans, Simon, and Simon Johnson. "The Comfort of Strangers." Swarm Toolkit. http://swarmtoolkit.net/index.php?option=com_content&task=view&id=18&Itemid=49.

2. E-mail interview with Simon Johnson, May 3, 2009.

3. Turner, Victor. *Dramas, Fields, and Metaphors: Symbolic Action in Human Society* (Ithaca: Cornell University Press, 1975), 45.

4. "Necklace of the Subaltern Betrayer." Ghosts of a Chance, September 15, 2008. http://ghostsofachance.com/main_site/index.php?p=object&id=1.

5. Bath, Georgina. "Ghosts of a Chance Alternate Reality Game Report." Smithsonian American Art Museum, Washington, D.C., November 6, 2008. http://ghostsofachance.com/GhostsofaChance_Report2.pdf.

6. Simon, Nina K. "An ARG at the Smithsonian: Games, Collections, and Ghosts." Museum 2.0, September 8, 2008. http://museumtwo.blogspot.com/2008/09/arg-at-smithsonian-games-collections.html.

7. Increased age is associated with negative qualities, such as decreases in stature, power, and cognitive ability, according to Mahzarin Banaji, who led the ageism studies at Harvard University. Published findings include Cunningham, W. A., M. K. Johnson, J. C. Gatenby, J. C. Gore, and M. R. Banaji. "Neural Components of Social Evaluation." *Journal of Personality and Social Psychology*, 2003, 85: 639–49.

CHAPTER 10

1. Saba, Moussavi, Somnath Chatterji, et al. "Depression, Chronic Diseases, and Decrements in Health: Results from the World Health Surveys." *The Lancet*, September 8, 2007, 370(9590): 851–58. DOI: 10.1016/S0140-6736(07)61415-9.

2. Lyubomirsky, *The How of Happiness*, 14.

3. Ibid., 7.

4. Ibid., 72.

5. Ibid.

6. Seligman, *Authentic Happiness*, xii.

7. Weiner, *The Geography of Bliss*, 310.

8. Ben-Shahar, *Happier*, 165–66.

9. Williams, Sam. "Hack, Hackers and Hacking." In *Free as in Freedom: Richard Stallman's Crusade for Free Software* (Sebastopol, CA: O'Reilly, 2002). Also available online at http://oreilly.com/openbook/freedom/appb.html.

10. Caplan, Jeremy. "Hacking Toward Happiness." *Time*, June 21, 2007. http://www.time.com/time/magazine/article/0,9171,1635844,00.html.

11. I coined the term in 2007, but I'd already been doing it for years by then. I just needed a way to describe it to other technologists so they could start doing it, too. I introduced the term in a keynote

at the Emerging Technology Conference (ETech) in San Diego, March 26–29, 2007, and again in a keynote at the Web 2.0 Expo in San Francisco, April 18, 2007.

12. Morrill, et al., *Together Alone*, 231.

13. This claim is based on my review of the findings of positive-psychology research trials on acts of kindness, in which participants typically don't know each other before being put in the cooperative test setting. It's a perfect example of transitory sociality.

14. Keltner, *Born to Be Good*, 3.

15. Ibid., 7–8.

16. These rules were coauthored with Sean Stewart, the award-winning science fiction and fantasy author, during the Last Call Poker project.

17. Tad Friend, "The Shroud of Marin." *The New Yorker*, August 29, 2005.

18. Davidson, Amy. "California Dying: Q&A with Tad Friend." *The New Yorker* online, August 29, 2005. http://www.newyorker.com/archive/2005/08/29/050829on_onlineonly01?currentPage=all.

19. Pujol, Rolando. "NYC Cemeteries Dying from Neglect." *AM New York*, May 29, 2009. http://amny.com/urbanite-1.812039/nyc-cemeteries-dying-from-neglect-1.1286733.

20. Weiner, *The Geography of Bliss*, 73.

21. Hecht, Jennifer Michael. *The Happiness Myth: The Historical Antidote to What Isn't Working Today* (New York: Harper One, 2008), 57.

22. Ibid., 58.

23. Ben-Shahar, *Happier*, 147–48.

24. Hecht, *The Happiness Myth*, 59.

25. Keltner, *Born to Be Good*, 195.

26. Quoted in Keltner, *Born to Be Good*, 186.

27. Hecht, *The Happiness Myth*, 298.

28. Csíkszentmihályi, *Beyond Boredom and Anxiety*, 102–21.

29. Keltner, *Born to Be Good*, 220.

30. Weiner, *The Geography of Bliss*, quoting John Stuart Mill, 74.

PART III

1. Flugelman, Andrew. "The Player Referee's Non-Rulebook." In *The New Games Book*, 86.

CHAPTER 11

Portions of this chapter are drawn from "Engagement Economy: The Future of Massively Scaled Participation and Collaboration," a *Technology Horizons* special report I created for the Institute for the Future in September 2008. The full original report, published by IFTF, is available at http://iftf.org/node/2306.

1. "How the *Telegraph* Investigation Exposed the MPs' Expenses Scandal Day by Day." *Telegraph*, May 15, 2009. http://www.telegraph.co.uk/news/newstopics/mps-expenses/5324582/How-the-Telegraph-investigation-exposed-the-MPs-expenses-scandal-day-by-day.html.

2. Wintour, Patrick, and Nicholas Watt. "MPs' Expenses: Critics Attack Censorship as Redactions Black Out Documents." *Guardian*, June 19, 2009. http://www.guardian.co.uk/politics/2009/jun/18/mps-expenses-censorship-black-out.

3. Jeff Howe. *Crowdsourcing: Why the Power of the Crowd Is Driving the Future of Business* (New York: Crown Business, 2008), 4–17.

4. Andersen, Michael. "Four Crowdsourcing Lessons from the *Guardian*'s (Spectacular) Expenses-Scandal Experiment." Nieman Journalism Lab, June 23, 2009. http://www.niemanlab.org/2009/06/four-crowdsourcing-lessons-from-the-guardians-spectacular-expenses-scandal-experiment/.

5. "Participation on Web 2.0 Websites Remains Weak." Reuters, April 18, 2007. http://www.reuters.com/article/technologyNews/idUSN1743638820070418.

6. Wright, Steven, and Jason Groves. "Shameless MPs Try to Dodge Trial Using 1689 Law Which Protects Them from Prosecution." *Daily Mail*, February 8, 2010. http://www.dailymail.co.uk/news/article-1248688/MPs-expenses-Three-Labour-MPs-Tory-peer-charged-false-accounting.html#ixzz0ikXMZse6.

7. Shirky, Clay. "Gin, Television, and Social Surplus." Personal blog, April 26, 2008. http://www.shirky.com/herecomeseverybody/2008/04/looking-for-the-mouse.html.

8. Internet World Stats—Usage and Population Statistics, accessed December 31, 2009. http://www.internetworldstats.com/stats.htm.

9. "Wikipedia Is an MMORPG." Wikipedia project, accessed May 1, 2010. http://en.wikipedia.org/wiki/Wikipedia:Wikipedia_is_an_MMORPG.

10. Ibid.

11. Puente, Maria. "Learn, Fight Hunger, Kill Time All at Once at Freerice.com." *USA Today*, January 23, 2008. http://www.usatoday.com/life/lifestyle/2008-01-23-freerice_N.htm.

12. The answer is "growing from the tip of a stem."

13. "Frequently Asked Questions." Free Rice. http://www.freerice.com/faq.html.

14. Anderson, Frank. "Can Videogames Make the World a Better Place?" BitMob, August 25, 2009. http://bitmob.com/index.php/mobfeed/foldinghome-distributed-computing.html.

15. Pietzsch, Joachim. "The Importance of Protein Folding." Horizon Symposia: Connecting Science to Life, online project of the journal *Nature*. http://www.nature.com/horizon/proteinfolding/background/importance.html.

16. McElroy, Griffin. "Joystiq Set to Overtake G4 in Folding@home Leaderboards Tonight." Joystiq, February 8, 2010. http://playstation.joystiq.com/tag/folding@home; and Stella, Shiva. "Sony Updates Folding@home for PS3 Folks Trying to Save the World." Game Bump, December 19, 2007. http://www.gamebump.com/go/sony_updates_folding_home_for_ps3_folks_trying_to_save_the_world.

17. Stasick, Ed. "Leave Your PS3 on for a Good Cause This Sunday Night!" Joystiq, March 21, 2007. http://playstation.joystiq.com/2007/03/21/leave-your-ps3-on-for-a-good-cause-this-sunday-night/.

18. Rimon, Noah. "Folding@home Petaflop Barrier Crossed." PlayStation blog, September 19, 2007. http://blog.us.playstation.com/2007/09/foldinghome-petaflop-barrier-crossed/.

19. Khan, Shaan. "Turning In-Game Achievements into Real-World Action." Thinkers & Doers, July 30, 2009. http://blogs.waggeneredstrom.com/thinkers-and-doers/2009/07/turning-in-game-achievements-in-real-world-action/.

20. Dumitrescu, Andrei. "PlayStation 3 Will Catch Up to the Xbox 360 in 2011." Softpedia, August 5, 2009. http://news.softpedia.com/news/PlayStation-3-Will-Catch-Up-to-the-Xbox-360-in-2011-118402.shtml.

21. Jacques, Robert. "Folding@home Clocks Up a Million PS3 Users." V3, February 6, 2008. http://www.v3.co.uk/vnunet/news/2208966/folding-home-clocks-million-ps3.

22. "Computer Game's High Score Could Earn Nobel in Medicine." RichardDawkins.net forum, May 11, 2008. http://forum.richarddawkins.net/viewtopic.php?f=5&t=44321.

23. The game was developed by doctoral student Seth Cooper and postdoctoral researcher Adrien Treuille, both in computer science and engineering, working with Zoran Popović, a University of Washington professor of computer science and engineering; David Baker, a UW professor of biochemistry and Howard Hughes Medical Institute investigator; and David Salesin, a UW professor of computer science and engineering. Professional game designers provided advice during the game's creation.

24. Fahey, Mike. "Humans Triumph Over Machines in Protein Folding Game Foldit." *Kotaku*. August 6, 2010. http://www.kotaku.com.au/2010/08/humans-triumph-over-machines-in-protein-folding-game-foldit/.

25. Cooper, Seth, Firas Khatib, et al., and Foldit players. "Predicting protein structures with a multiplayer online game." *Nature*. August 2010. 466, 756–60. http://www.nature.com/nature/journal/v466/n7307/full/nature09304.html.

26. Deci, Edward L., Richard Koestner, and Richard M. Ryan. "A Meta-Analytic Review of Experiments

Examining the Effects of Extrinsic Rewards on Intrinsic Motivation." *Psychological Bulletin*, November 1999, 125(6): 627–68. DOI: 10.1037/0033-2909.125.6.627.

27. "Job Listings." Bungie.net, accessed May 1, 2010. http://www.bungie.net/Inside/jobs.aspx.

CHAPTER 12

1. "Epic Win." Urban Dictionary entry. http://www.urbandictionary.com/define.php?term=epic%20win.

2. Twitter replies to @avantgame from @tobybarnes, @changeist, and @incobalt, on November 4, 2009.

3. Many of The Extraordinaries' microvolunteer missions are more like Amazon Mechanical Turk–style human intelligence tasks — labeling, tagging, and sorting digital objects for museums, scientists, and government agencies, for example. This is important, but it doesn't represent the epic win. In phone conversations with the founders of the project in early 2010, I learned that the company aims to focus on developing more real-world, mobile missions and rely less on online human intelligence tasks.

4. "Chat with an Extraordinary: Nathan Hand of Christel House." The Extraordinaries blog, November 17, 2009. http://www.theextraordinaries.org/2009/11/chat-with-an-extraordinary-nathan-hand-of-chris tel-house.html.

5. Wenner, Melissa. "Smile! It Could Make You Happier." *Scientific American*, September 2009. http://www.scientificamerican.com/article.cfm?id=smile-it-could-make-you-happier.

6. "Chat with an Extraordinary," The Extraordinaries blog.

7. Official Game Guide for *The Sims* 3. Prima Official Game Guides, June 2, 2009, 42. http://www.primagames.com/features/sims3/.

8. Edelman, Joe. "Make a Wish." Live talk, Ignite Amherst, September 22, 2009. http://igniteshow.com/amherst.

9. Personal interview with Joe Edelman, July 12, 2008.

10. Edelman, "Make a Wish."

11. Ibid.

12. Edelman, Joe. "Your Life: The Groundcrew Mission." Groundcrew. http://groundcrew.us/papers/your-life.

13. Edelman, Joe. "The Mobile Manifesto: How Mobile Phones Can Replace a Broken Economy." Joe Edelman blog, January 2009. http://nxhx.org/thoughts/manifesto.html.

14. Edelman, Joe. "Volunteering with Groundcrew." Groundcrew. http://groundcrew.us/papers/volun teering.

15. Jeyes, Dave. "Google Wants to Smarten Up Your Home with PowerMeter." Tech.Blorge, February 10, 2009. http://tech.blorge.com/Structure:%20/2009/02/10/google-wants-to-smarten-up-your-home-with-powermeter-2/.

16. Kho, Jennifer. "Adaptive Meter: Playing the Energy Conservation Game." Earth2Tech, March 29, 2009. http://earth2tech.com/2009/03/29/playing-the-energy-conservation-game/.

CHAPTER 13

1. Richards, C. "Teach the World to Twitch: An Interview with Marc Prensky." Futurelab, December 2003. www.futurelab.org.uk/resources/publications_reports_articles/web_articles/Web_Article578; and "Designing Games with a Purpose." *Communications of the ACM*, August 2008, 51(8): 58–67. DOI: http://doi.acm.org/10.1145/1378704.1378719.

2. Tomasello, Michael. *Why We Cooperate* (Cambridge: MIT Press, 2009).

3. Tomasello, Michael, and Malinda Carpenter. "Shared Intentionality." *Developmental Science*, December 2006, 10(1): 121–125.

4. Rakoczy, Hannes, Felix Warneken, and Michael Tomasello. "The Sources of Normativity: Young Children's Awareness of the Normative Structure of Games." *Journal of Developmental Psychology*, May 2008, 44(3): 875–81.

5. "Which Do You Prefer: Competitive or Cooperative Multiplayer?" *Escapist*, October 24, 2009. http://www.escapistmagazine.com/forums/read/9.151529.

6. Contreras, Paul Michael. "LittleBigPlanet Sets Another Milestone for Number of Levels." PlayStation Lifestyle, November 20, 2009. http://playstationlifestyle.net/2009/11/20/littlebigplanet-sets-another-milestone-for-number-of-levels/.

7. Brunelli, Richard. "Grand Prize Winner: McDonald's Brave New World." *AdWeek*, December 1, 2008. http://www.adweek.com/aw/content_display/custom-reports/buzzawards/e3i9417c5a4a703467 d97b51be9e35149f8.

8. The term was coined by Sean Stacey, who runs Unfiction (www.unfiction.com), the leading community forum since 2003 for chaotic fiction and cross-media alternate reality games.

9. Peterson, Christopher, and Martin Seligman. *Character Strengths and Virtues: A Handbook and Classification* (New York: Oxford, 2004).

10. "The VIA Classification of Character Strengths." The Values in Action (VIA) Institute, October 23, 2008. http://www.viacharacter.org/Classification/tabid/56/Default.aspx.

11. Tapscott, Don, and Anthony D. Williams. *Wikinomics: How Mass Collaboration Changes Everything* (New York: Portfolio, 2008), 33.

12. Ibid., x.

CHAPTER 14

1. Brand, Stewart. *Whole Earth Discipline: An Ecopragmatist Manifesto* (New York: Viking, 2009), 275, 298.

2. Brand, Stewart. "The Purpose of the *Whole Earth Catalog*." *Whole Earth Catalog*, Fall 1968. Electronic version available at http://wholeearth.com/issue/1010/article/195/we.are.as.gods.

3. Brand, *Whole Earth Discipline*, 276.

4. Ibid., 298.

5. Ibid.

6. Ibid., 276.

7. "Jill Tarter and Will Wright in Conversation." *Seed*, September 2, 2008. http://seedmagazine.com/content/article/seed_salon_jill_tarter_will_wright/.

8. "Spore." Wikipedia entry, accessed May 1, 2010. http://en.wikipedia.org/wiki/Spore.

9. For a history of the term "massively multiplayer forecasting games" and their development at the Institute for the Future, see: "Institute for the Future Announces First Massively Multiplayer Forecasting Platform." Institute for the Future. Palo Alto: September 22, 2008. http://iftf.org/node/2319; "Massively Multiplayer Forecasting Games: Making the Future Real." Institute for the Future. Palo Alto: September 7, 2008. http://iftf.org/node/2302.

10. Simon, Nina. "The Aftermath of the ARG World Without Oil." Museum 2.0, July 27, 2007. http://museumtwo.blogspot.com/2007/07/game-friday-aftermath-of-arg-world.html.

11. Cerulo, Karen. *Never Saw It Coming: Cultural Challenges to Envisioning the Worst* (Chicago: University of Chicago Press, 2006).

12. Gravois, John. "Think Negative!" Slate, May 16, 2007. http://www.slate.com/id/2166211.

13. "A to Z: A World Beyond Oil." World Without Oil, May 31, 2007. http://community.livejournal.com/worldwithoutoil/20306.html.

14. Guité, François. Quoted on "Buzz: World Without Oil," World Without Oil. http://www.worldwithoutoil.org/metabuzz.htm.

15. "Everything Falls Apart—the End (Semi-OOG)." WWO player blog post, May 31, 2007. http://fallingintosin.livejournal.com/12325.html.

16. "Fond and Sad Goodbye." World Without Oil—The Texts, June 1, 2007. http://wwotext.blogspot.com/2007/06/fond-and-sad-goodbye.html.

17. "Ending Thoughts (OOG)." WWO player blog post, May 31, 2007. http://monkeywithoutoil.blogspot.com/2007/05/ending-thoughts-oog.html.

18. "New Chess Theory Not for Einstein: Scientist Denies Ever Playing 'Three-Dimensional' Game, Even for Relaxation." *New York Times*, March 28, 1936.

19. Ibid.

20. Reid, Raymond. "Chesmayne: History of Chess." 1994. Accessible at http://www.chess-poster.com/english/chesmayne/chesmayne.htm.

21. Cascio, Jamais. "Super-Empowered Hopeful Individual." Open the Future, March 2008. http://openthefuture.com/2008/03/superempowered_hopeful_individ.html.

22. Friedman, Thomas. *The Lexus and the Olive Tree: Understanding Globalization* (New York: Farrar, Straus and Giroux, 1999), 381.

23. Ibid.

24. This particular saying is taken from "The Fifty-Year View," Jamais Cascio's April 2009 Ten-Year Forecast presentation for the Institute for the Future. http://iftf.org/files/deliverables/50YearScenarios.ppt.

25. "Superstruct FAQ." Institute for the Future, September 22, 2008. http://www.iftf.org/node/2096.

26. Demographic analysis by Kathi Vian for the 2009 Ten-Year Forecast report, presented at the annual retreat. http://www.iftf.org/node/2762.

27. The Institute for the Future. www.iftf.org.

28. Hersman, Erik. "Solving Everyday Problems with African Ingenuity." Africa Good News, February 2009. http://www.africagoodnews.com/innovation/solving-everyday-problems-with-african-ingenuity.html.

29. Becker, Gene. "The Long Game." Superstruct, 2008. http://www.superstructgame.org/Superstruct View/510.

30. Internet World Stats—Usage and Population Statistics.

CONCLUSION

1. Yudkowsky, Eliezer. "If You Demand Magic, Magic Won't Help." Less Wrong, March 22, 2008. http://lesswrong.com/lw/ou/if_you_demand_magic_magic_wont_help/.

2. That's 5 million years dating back to the start of the hominid species timeline, at *Australopithecus ramidus*.

3. "Archaeological Tree-Ring Dating at the Millennium." *Journal of Archaeological Research*, September 2002, 10(3): 243–75. http://www.springerlink.com/content/hhv8qd78hh7r5pfb/.

4. "Ancient Etruscans Were Immigrants from Anatolia, or What Is Now Turkey." European Society of Human Genetics, press release, June 16, 2007. http://www.eurekalert.org/pub_releases/2007-06/esoh-aew061307.php; see also Wade, Nicholas. "DNA Boosts Herodotus' Account of Etruscans as Migrants to Italy," *New York Times*, April 3, 2007. http://www.nytimes.com/2007/04/03/science/03etruscan.html.

APPENDIX 2

1. Report by the Mental Health Advisory Team for the Office of the Command Surgeon, US Forces Afghanistan and the Office of the Surgeon General, United States Army Medical Command. November 6, 2009. The video gaming data is in Section 6.2 "Individual Coping Behavior" on pages 33–34. Available online: http://www.armymedicine.army.mil/reports/mhat/mhat_vi/MHAT_VI-OEF_Redacted.pdf.

2. Gentile, Douglas, et al. "Pathological Video Game Use Among Youths: A Two-Year Longitudinal Study." *Pediatrics*. Published online January 17, 2011. http://pediatrics.aappublications.org/content/early/2011/01/17/peds.2010-1353.abstract.

3. Holt-Lunstad, Julianne, Timothy B. Smith, and J. Bradley Layton. "Social Relationships and Mortality Risk: A Meta-analytic Review." *Public Library of Science*. July 2010. Available online: http://www.plosmedicine.org/article/info%3Adoi%2F10.1371%2Fjournal.pmed.1000316. Lyubomirsky, Sonja,

Laura King, Ed Deiner. "The Benefits of Frequent Positive Affect: Does Happiness Lead to Success?" *Psychological Bulletin*. American Psychological Association 2005, 131 (6): 803–55 doi: 10.1037/0033-2909.131.6.803, Available online: http://www.faculty.ucr.edu/~sonja/papers/LKD 2005.pdf.

4. Coyne, Sarah M., Laura M. Padilla-Walker, Laura Stockdale, and Randal D. Day. "Game On . . . Girls: Associations Between Co-playing Video Games and Adolescent Behavioral and Family Outcomes." *Journal of Adolescent Health*, January 2011. doi:10.1016/j.jadohealth.2010.11.249. Available online: http://jahonline.org/webfiles/images/journals/jah/feature3.pdf.

5. Oxford, Jonathan, Davidé Ponzi, David C. Geary. "Hormonal Responses Differ When Playing Violent Video Games Against an Ingroup and Outgroup." *Evolution and Human Behavior* 31 (3): 201–9 (May 2010) doi: 10.1016/j.evolhumbehav.2009.07.002. Available online: http://www.ehbonline.org/article/S1090-5138(09)00067-1/abstract.

Index

Page numbers in *italics* refer to illustrations.

www.vintage-books.co.uk